Quantum Space-Time Dynamics

A Theory of Space-Time and Gravity at the smallest distance where measurement still makes sense.

By

عائشة عبد الرشيد
Aisha Abd Ar-Rashid
(Hontas Farmer)

© Hontas Farmer 2009 All rights reserved.
Bellwood Scientific Press
Bellwood, IL, Chicago, IL
United States of America
ISBN:978-0-578-00732-8

Table of Contents

Foreword

Who am I ? My legal name is Hontas Farmer however I like to be called by my arabic name Aisha Abd Ar-Rashid. I am realistic about being called that though. How long has it been since Cassius Clay became Mohammed Ali, and is he not still called Clay by some? So don't sweat the name. I am a theoretical physicist from Chicago educated at Northern Illinois University for my Bachelors, I practically have a M.S. from the University of Illinois at Chicago. It was at UIC that I was educated formally in graduate level quantum mechanics, quantum field theory, theoretical particle physics, and general relativity. It was at UIC where I did the basic theoretical research that went into this theory. I was inspired to research quantum gravity on my very last day at Northern Illinois University. I was in a student seminar that was required before graduation. One students topic was gravity and the subject of quantum gravity came up. A professor sitting next to me said in response to a question "There is no accepted theory of quantum gravity". I thought to myself wow, well this is a wide open field then. That is why I started down this road. Due to a disagreement with some people at UIC which I will not air out here, I find myself at Depaul University in Chicago. So bad was the dispute that I was unable to get even the letters of recommendation I needed to get into most institutions. *Depaul admitted me solely on the strength of my transcript.* It is from there that god willing I will get a MS and in short order admission to a PhD. Program. This to me is just a formality. For with the publication of this book, I feel that I have done the equivalent of a PhD. Such an unorthodox, way of achieving success is not uncommon for people like me. For you see I am a male to female transsexual. I think that a part of my difficulty at UIC was due to this fact. (I don't think that Depaul knew that). There are some people who just cannot see a person like me being a physicist even when they themselves have given me A's and B's in the subject.

Why have I written such a book and decided to publish it via Lulu and online instead of via a journal? Well that's easy because I could not get even parts of the work published in a mainstream scientific journal. The referee's basically could not understand what I wrote here. In particular they could not understand my proof regarding the field with the same symmetry as the Lie group and Algebra F(4)/Spin(4). They could not say it was "wrong" but they could not say it was "right". I attempted to publish this book via the arXiv archive. Based on my institutional affiliation it was at first scheduled for broadcast to the physics community then it was suddenly withdrawn a the 11th hour. With no real explanation. I know I run a great risk of sounding like a crank, but whatever the "giggle factor" is what it is. So giggle away. All I ask is that any serious student of quantum gravity critically read my inexpensive little book, think about it, and keep it in mind. You may giggle all you want now, however maybe just maybe at some point in the future when more direct experiments can be done to confirm or repudiate the theory herein you will be a step ahead.

There is one thing that could happen which would disprove this theory; *Any discovery of so called "large extra dimensions".* This theory predicts that any extra dimensions if they exist MUST be of Planck Scale or smaller.
 Large extra dimensions are something that many prominent scientist want to have. Some formulations of string theory, and M theory would get a real boost from such a discovery. In the postulation of my theory I am implying that such large extra dimensions cannot exist. If they are found then this theory is false.
 Detection of a gravitational attraction by

There are a number of things that would not falsify this theory however.

The discovery of heavier particles (which may or may not be super symmetric).

The discovery of a new force or other non-gravitational extension to the standard model.

The way this theory is structured it can accommodate any non gravitational action, and solve for the quantum gravitational field of any configuration of particles and fields imaginable.

There are some experiments which could verify parts of this theory.

The basic premise is that space-time is discrete. So any experiment that confirms that space-time is discrete will verify one large part of this theory.

This theory predicts that a photon with a wavelength equal to the Planck length will collapse to form a quantum black hole.

This theory makes a negative prediction, masses smaller than the planck mass will have NO gravitational field of their own. Unless they are somehow bound together non-gravitationally. (This flies in the face of the dark matter paradigm which has most of space-filled with particles of much less mass than the Planck mass which only interact by gravity. While if dark matter were directly detected it would not disprove this theory. It is possible the dark matter interacts by some other non gravitational force which we don't know or understand, and forms clumps that are larger than one Planck mass.) A null result on detecting the gravity from any quantum particle would be a confirmation.

In the process of publishing this book I realize that full justice cannot be done to it except by downloading the mathematica notebook, and wolfram's mathreader and seeing it in action. I will keep the mathematica notebook source file for this book on my personal website http://www.geocities.com/hontasfx for as long as a yahoo-geocities website is still free. If that is no longer the case i suggest googleing either Hontas Farmer, or my recently adopted Arabic name Aisha Abd Ar-Rashid.

Quantum Space-Time Dynamics

A theory of Space-Time and Gravity at the smallest distance where measurement still makes sense.

Hontas F. Farmer (Aisha Abd Ar-Rashid)

University of Illinois at Chicago, DePaul University

Physics Departments

Keywords: Quantum gravity, Planck Scale, F(4)/Spin(4), Neutron

PACS: 03.70.+k 04.60.Gw 04.70.Dy 04.90.+e

The glaring theoretical problem for 21 st century physics is how to unite the two great theories of 20 th century physics. I have done just this by considering the problem from a fresh new perspective. Giving due acknowledgement to the established contenders. I must admit that when I started my research I knew nothing of them and so with a fresh unfettered mind was able to find the right answer. This theory predicts the effects of gravity at the quantum scale, the classical scale, and introduces corrections at the cosmological scale that have implications for theories of alternative gravity and/or dark matter. The theory reproduces the expected result for the entropy of a quantum black hole and allows a thermodynamic description of their physics. The theory can be used to model an actual experimental observation of the behavior of ultra cold neutrons in earth' s gravitational field. When put into a form familiar to those who know Quantum Field Theory the theories F (4)/Spin (4) symmetry and Lie algebra become obvious. Which by Noether' s theorem leads to the conclusion that space - time is a conserved quantity (it has not been created or destroyed since the big bang). Then I make connections to concrete observations that have been made by NesvizhevskyAll of this is derived from a single principle, the postulate of quantized space - time; Mass - energy alters the local space - time interval in increments that are integer multiples of the Planck length. This is the compilation of works done separately over 4 years. Originally composed using Mathematica.

Review of Current Research in the Literature.

Gravitation is the most fundamental interaction in the universe. Absolutely everything interacts gravitationally with everything else. Currently we have no internally consistent, experimentally verifiable, and agreed upon fundamental quantum theory of gravity. There are several contenders for such a theory they all fail in one of the mentioned respects. Notable among them is the theory of Loop Quantum gravity and String/(M)atrix theory, and Lattice Quantum gravity.

String/M-Theory is popular and well publicized. Many many physicist believe in it. String theory was first proposed by Yoschiro Nambu. He was considering the problem of Quark confinement. From there string physics evolved into many different theories bosonic strings, heterotic strings, super symmetric strings, etc, etc. All of these variants were mathematically consistent. This is a problem for a fundamental theory of everything. M-Theory solves this problem by proposing a different fundamental object called a brane. There is only one fundamental M theory. String/M-theory claims to be a theory of all possible interactions and can postdict fundamental constants. So far those sets of fundamental constants or vacuum states are not unique. Estimates of the number of vacuums are as high as ten to the five hundreth power possible vacuums. In light of this string/M theorist propose a cosmological landscape of vacuum states and our vacuum state is only one of those. This is a problem because there is no way even in principle that this claim can ever be verified.

Loop Quantum gravity is a canonical theory of quantization of the gravitational field. It employ's Ashtekar variables to rewrite Einstein equations then uses loop quantization to second quantize the field. This theory contains one free parameter called the Immirzi parameter. This free constant plays a critical role in the crucial computation of the enthropy of a black hole. The computation of the enthropy of a black hole by LQG is only correct if the parameter is chosen to make it correct. This presents a problem because a fundamental part of the theory needs to be found from classical theory. In a sense it makes the enthropy test in applicable to Loop Quantum gravity. Unlike string/M theory Loop quantum gravity is just a theory of gravitation and nothing more. Loop quantum gravity proposes that space-time is at the finest possible scale a spinfoam. This spinfoam is a dynamically evolving quantum foam that only looks sooth from a great distance. I do not like this feature as there is no way to verify that there is a spinfoam as opposed to a more generic lattice.

Euclidian/Lorentzian lattice quantum gravity hypotheses that space-time is a lattice with lattice constant equal to the Planck length. These theories are also theories of canonical quantum gravity. These theories further postulate that the space-time lattice has a certain geometry simple cubic,hexagonal etc. These different lattice shapes seem to set a preferred coordinate system. Gravity is proposed to emerge from the interaction of mass-momentum currents through this lattice. Gravitons arise as phonons in the lattice. The problem I have with this theory is that it makes assumptions about the exact structure of discrete space-time that cannot be elucidated by experiment. That is to say every experiment I have seen proposed would only prove that space-time is discrete and not that a specific theory of discrete space-time was correct.

The theory I am proposing is based on a single hypothesis I call the postulate of quantized space-time. The result is a theory of quantum geometry and space-time. This theory will not have the problem of violating Lorentz invariance of the speed of light. This theory will not have the problem of making predictions that cannot be substantiated by experiment. Specifically I will derive a fully quantum mechanical formula for Schwarzschild radius of a black hole, as well as it's entrhopy. The calculation will find a result which accords with that of Beckenstein and Hawking.

Observations

The Gravitational force seems to pervade the entire universe. However gravity is the weakest of all the forces in the entire universe. To get a easily measurable force of gravity there needs to be a planetary or larger mass nearby. So any theory of quantum gravity must require a large amount of mass to create a small effect in space-time. Gravity also seems to be an exclusively attractive force. It also has only monopole and quadrupole moments. The lowest radiative gravitational moment is the quadrupole moment.

There is no way to distinguish between a mass and bound energy state. Mass is energy in a bound state. Only concentrated in these bound states called mass have we actually observed the gravitational field of an object. Free energy seems to react to gravitational fields but does not possess a gravitational field of it's own. Furthermore the bound states of energy are quantized into discrete steps.

The Planck length seems to be the smallest possible interval of length in space. The Planck time seems to be the smallest physically meaningful division of time. Furthermore the invariant interval computed by combining the Planck interval with the planck time seems to be the smallest physically meaningful interval between events in space-time. If we choose units such that $\hbar = c = 1$ (Such will be the system of units throughout this paper.) the Planck units then take on the following values. $t_p = \sqrt{G}$, $l_p = \sqrt{G}$, $m_p = \frac{1}{\sqrt{G}}$ note $t_p = l_p = \frac{1}{m_p}$

In quantum physics all particles and fields seem to have a intrinsic angular momentum called spin. Spin angular momentum must be an intrinsic part of the quantized space-time. In quantum field theory it is customary to refer to a spin zero bosonic field as a scalar field, and a spin 1 bosonic field as a vector field. For some reason with gravity and it's proposed graviton of spin-2 it is also customary to refer to it as a vector field. Because it vectors or carries a fundamental force. I will break with this idea and call the field a tensor field. The spin of a particle or field in this theory will be reflected by the tensor rank of it's mathematical representation. In the case of fermions the field will generally be represented by a vector.

Gravitation will serve to alter the probabilities of the interactions in a given space as compared to the probabilities when the space is presumed flat.

From the old quantum theory we have the formula due to DeBroglie and Einstein which says that Energy / Momentum $= \frac{1}{\text{wavelength}}$. Keep this in mind.

Suppose we compute the Schwarzschild radius of a Black Hole with mass equal to one Planck mass.

$$R_{Schwarzschild} \, (1 \, m_p) = 2 \, \frac{1}{\sqrt{G}} \, G = 2 \, \sqrt{G} \tag{0.1}$$

Now suppose we compute the invariant interval between two points separated by one Planck unit in 3-space and time.

$$\sqrt{\left(-\sqrt{G}\right)^2 + \left(\sqrt{G}\right)^2 + \left(\sqrt{G}\right)^2 + \left(\sqrt{G}\right)^2} = 2 \, \sqrt{G} \tag{0.2}$$

Hmmm. Could this be a pure coincidence?

The Postulate of Quantized Space-Time

Mass-energy alters the local space-time interval in increments that are integer multiples of the Planck length.

There have been multiple attempts and advocacies for using the Planck units as a fundamental system of units. I am not looking to advocate for that. What I do claim is that by breaking length-time into intervals that are integers times the Planck length in just the right way a theory of dynamical quantum space-time and hence gravitation can be formulated. What are the consequences of this postulate? To answer this I will translate these words into mathematical expressions.

First I will define the space that this theory lives in and call it the space of physical vectors. First I will prove this space supports calculus, that it is a differentiable manifold. Then I will define its metric and show that all possible metrics for the space of physical vectors are complete. Thus showing that all existing solutions are Hilbert spaces. Then I will define operators on this space of physical vectors.

The Space of Physical Vectors

Before the dynamics of quantum space-time can be intelligently discussed we must define exactly what space-time is. In this theory space-time is referred to as the space of physical vectors. The @ operator matrix is a form of metric tensor derived from the Minkowski metric. Along with the definition of the displacement operators of the form $\triangle \hat{x}$ defines a set of vectors and a metric. This set of vectors along with this metric will define an inner product space If it is closed under addition and scalar multiplication.

Proof that the set of vectors with the form x^μ , defined below, is a inner product space.

Let $\gamma \epsilon \quad c$

$$x^\mu = \sqrt{G} \ (-\gamma \ n_0, \ \gamma \ n_1, \ \gamma \ n_2, \ \gamma \ n_3) \tag{0.3}$$

This is just a vector composed of quantum numbers. As such γ can be factored out.

$$\Longrightarrow \gamma \sqrt{G} \ (-n_0, \ n_1, \ n_2, \ n_3) \tag{0.4}$$

Therefore the set is closed under scalar multiplication. As for linear combination.

$$x^\mu + x^\nu \Longrightarrow \sqrt{G} \ (-n_0, \ n_1, \ n_2, \ n_3) + \sqrt{G} \ (-m_0, \ m_1, \ m_2, \ m_3) \tag{0.5}$$

$$\Longrightarrow \sqrt{G} \ ((-n_0, \ n_1, \ n_2, \ n_3) + (-m_0, \ m_1, \ m_2, \ m_3)) \tag{0.6}$$

$$\Longrightarrow \sqrt{G} \ (-n_0 - m_0, \ n_1 + m_1, \ n_2 + m_2, \ n_3 + m_3) \tag{0.7}$$

Let $n_i + m_i = k_i$

$$\Longrightarrow \sqrt{G} \ (-k_0, \ k_1, \ k_2, \ k_3) \tag{0.8}$$

Which is of the same form as the two vectors input. Therefore the set is closed under linear combination. Therefore this is a vector space. Q.E.D

The inner product and metrics

The next crucial question is what is the inner product on this vector space? What is the metric on this space-time? I propose this definition... $\chi_{\mu\nu}$ is a dynamical metric tensor/operator with coefficients $m_i \epsilon \ \mathbb{Z} \geq 1$. These coefficients will be called "coefficients of geometry" and can be determined from the modified Hilbert stress tensor cited below (equation #).

$$\text{In[1]:= } \hat{\chi} = \begin{pmatrix} -n_0 & 0 & 0 & 0 \\ 0 & n_1 & 0 & 0 \\ 0 & 0 & n_2 & 0 \\ 0 & 0 & 0 & n_3 \end{pmatrix}$$

Out[1]= $\{\{-n_0, 0, 0, 0\}, \{0, n_1, 0, 0\}, \{0, 0, n_2, 0\}, \{0, 0, 0, n_3\}\}$

Using this operator as the most general metric tensor I can define the inner product of two position vectors U and V.

In[83]:= $U = \begin{pmatrix} -u_0 \sqrt{G} \\ u_1 \sqrt{G} \\ u_2 \sqrt{G} \\ u_3 \sqrt{G} \end{pmatrix}$

In[84]:= $V = \begin{pmatrix} -v_0 \sqrt{G} \\ v_1 \sqrt{G} \\ v_2 \sqrt{G} \\ v_3 \sqrt{G} \end{pmatrix}$

In[90]:= $\langle U \mid V \rangle$ = **Transpose[U]** $.\hat{\chi}.$ **V**

By examination of this output I see that if the m_i equal one then this metric becomes the Minkowski metric. Like the Minkowski metric all metrics of this form will have convergent Cauchy sequences and so be complete metrics. Therefore making the space of physical vectors a Hilbert space

I claim :

At the Planck scale all possible metrics are equivalent.

Proof:
Consider the minimum interval between two points p and q in space-time given the postulate of quantized space-time. This will be substituted for the differential of path length. Use the inner product and metric as defined above.

In[97]:= $P = \begin{pmatrix} -1 \sqrt{G} \\ 1 \sqrt{G} \\ 1 \sqrt{G} \\ 1 \sqrt{G} \end{pmatrix}$

In[95]:= $q = \begin{pmatrix} -1 \sqrt{G} \\ 1 \sqrt{G} \\ 1 \sqrt{G} \\ 1 \sqrt{G} \end{pmatrix}$

In[91]:= $\Delta S = \sqrt{\langle p \mid q \rangle} =$

In[98]:= $\sqrt{\textbf{Transpose[p]}.\hat{\chi}.\textbf{q}}$

Out[98]= $\left\{ \left\{ \sqrt{-G\, m_0 + G\, m_1 + G\, m_2 + G\, m_3} \right\} \right\}$

Substitute this into the equation for the geodetic line as given in section 9 equation 20 of [7]. For now convert the integral into a Riemann sum.

$$\int_a^b dS = \int_a^b \Delta S =$$

$$\text{Limit}\left[\left[\sum_{m_1=1}^{n} \sqrt{-G\, m_0 + G\, m_1 + G\, m_2 + G\, m_3} \right], \text{max}\Delta S \to 0 \right] = \sqrt{2\, n\, G} =. \quad \text{Where } n = \frac{b-a}{\sqrt{G}}.$$

The postulate of quantized space-time asserts that the differentials in fact are of the Planck length. Therefore at the Planck scale and close to it all solutions to Einstein's equations will converge to this minimal interval.
QED.

Another way of putting this is that at the Planck scale all metrics would converge to the Minkowski metric. Between two such narrowly separated points only special relativity is in effect. This is in agreement with the statement by Einstein in section four paragraph two of [7]. Where Einstein writes "For infinitely small four-dimensional regions the theory of relativity in the restricted sense (special relativity) is appropriate, if the coordinates are suitably chosen." Thus this part of the theory is in agreement with the assumptions made by Einstein.

Does this space support calculus?

How about calculus? Are functions on this space-time differentiable? Integrable? Yes, and yes, here is proof. Differentiation is by simple definition..

$$\frac{d}{dx} f[x] = \lim_{\Delta x \to 0} \frac{f[x + \Delta x] - f[x]}{\Delta x} = \text{finite quantity} \tag{0.9}$$

Defined if the limit is finite.

In the space defined the change in x is and can be no smaller than $n\sqrt{G}$.

$$\lim_{\Delta x \to 0} \frac{f[x + \Delta x] - f[x]}{\Delta x} \Longrightarrow \lim_{n \to 1} \frac{f[n\sqrt{G} + n\sqrt{G}] - f[n\sqrt{G}]}{n\sqrt{G}} \tag{0.10}$$

$$\Longrightarrow \frac{f[n\sqrt{G} + \sqrt{G}] - f[n\sqrt{G}]}{\sqrt{G}} \tag{0.11}$$

The denominator is larger than zero and the numerator is larger than zero. Therefore the limit will generally evaluate to a finite quantity. Therefore the derivative exist. Therefore the manifold as defined is differentiable and supports calculus. Q.E.D.

Though derivatives and integrals exist on this manifold they may not be exactly the same as they are in smooth space-time. For example the derivative of x^2.

$$\lim_{\Delta x \to 0} \frac{[x + \Delta x]^2 - x^2}{\Delta x} \Longrightarrow \lim_{n \to 1} \frac{[n\sqrt{G} + \sqrt{G}]^2 - (n\sqrt{G})^2}{\sqrt{G}} \tag{0.12}$$

$$\Longrightarrow \lim_{n \to 1} \frac{(n\sqrt{G})^2 + 2n\sqrt{G}\sqrt{G} + \sqrt{G}^2 - (n\sqrt{G})^2}{\sqrt{G}} = \tag{0.13}$$

$$\frac{\sqrt{G} + \sqrt{G}^2}{\sqrt{G}} = \lim_{n \to 1} \frac{\sqrt{G}(2n\sqrt{G} + \sqrt{G})}{\sqrt{G}} = \lim_{n \to 1} 2n\sqrt{G} + \sqrt{G} = 3\sqrt{G}$$

Translated back into the usual notation $x \to n\sqrt{G}$ ..

$$= \lim_{n \to 1} 2n\sqrt{G} + \sqrt{G} = 2x + \sqrt{G} \tag{0.14}$$

So the difference between a derivative evaluated the traditional way and one evaluated on this quantized space-time will be on the same order of magnitude as the Planck length. Furthermore this additive constant will not carry over through subsequent differentiations. For most purposes this little uncertainty could be safely ignored in all but the finest calculations.

I will define the quantum derivative Δ using equation 14 as a guide. In this composition I will ignore the planck scale error and juts have mathematica carry out the usual derivative. This will not introduce much error to the calculations I am going to carry out. Say f[x] is a linear function.

$$\frac{\Delta f[x]}{\Delta x} = \lim_{\Delta x \to 0} \frac{f[x + \Delta x] - f[x]}{\Delta x} =$$

$$\frac{f[n\sqrt{G} + \sqrt{G}] - f[n\sqrt{G}]}{\sqrt{G}} = \frac{f[n\sqrt{G}] + f[\sqrt{G}] - f[n\sqrt{G}]}{\sqrt{G}} = \frac{f[\sqrt{G}]}{\sqrt{G}}$$

Base on the above without making any assumptions about the properties of "f" I can define the derivative on this space-time, the quantum derivative, in terns of the derivative on smooth space-time as the following expression.

$$\frac{\Delta f[x]}{\Delta x} = \frac{d f[x]}{dx} + \sqrt{G} \tag{0.15}$$

How about integration in this space-time? Integration has to be defined in such a way so that it will invert differentiation. I know that in the traditional way of defining the integral and the derivative this is not taken into consideration. The inverse relationship between integration and differentiation is sort of an accident. Here it will be quite on purpose as I want the results of solving a differential equation to be the same on this space as they would be on a smooth classical space-time. With the last equation in mind I define the Riemann integral on this space-time in terns of the Riemann integral of f[x] on smooth space-time like so.

$$\int f[x] \; \Delta \; x = \int \left[f[x] - \sqrt{G} \; \right] dx \qquad (0.16)$$

With definition the result of taking a derivative then and integral or an integral then a derivative on this quantized space-time gives the exact same results as it would in smooth space-time. Thus the solutions to differential equations remain the same so no laws of physics already established need to be altered for the sake of this theory.

$$\int \frac{\Delta \; f[x]}{\Delta \; x} \; \Delta \; x = \int \left[\left[\frac{d \; f[x]}{dx} + \sqrt{G} \; \right] - \sqrt{G} \; \right] dx = \int \left[\frac{d \; f[x]}{dx} \right] dx = f[x] \qquad (0.17)$$

I have accomplished in this section what I promised. I have shown that the space of physical vectors is a differentiable and integrable manifold of Hilbert space regardless of the exact values of elements of $\hat{\chi}$. I have further defined the calculus on this space in such a way as to preserve the solutions to all known differential operators. I have show that the difference between a derivative or integral in the space of physical vectors is exactly one planck interval. $\sqrt{G}$. The stage is now set to define proper differential operators on this space-time.

Definitions of the operators and their algebra.

Before the postulate of quantized space-time is given mathematical realization definitions are in order. I will define the operators for the benefit of the reader and of *Mathematica*. So I can just have the computer evaluate the expressions for me. Thus sparing me any stupid mathematical errors. First I will define the position and momentum operators in a way that *Mathematica* can work with.

$$\text{Let } n_i \; |_{i \in \{0,1,2,3\}} \; \text{ be positive integers.} \qquad (0.18)$$

The position vector/ operator will be defined as.

In[11]:= $\mathbf{x}^{\mu} = \sqrt{G} \; (- \; n_0, \; n_1, \; n_2, \; n_3)$

In[2]:= $\mathbf{X} = \begin{pmatrix} - n_0 \sqrt{G} \\ n_1 \sqrt{G} \\ n_2 \sqrt{G} \\ n_3 \sqrt{G} \end{pmatrix}$

Out[2]= $\left\{ \left\{ - \sqrt{G} \; n_0 \right\}, \; \left\{ \sqrt{G} \; n_1 \right\}, \; \left\{ \sqrt{G} \; n_2 \right\}, \; \left\{ \sqrt{G} \; n_3 \right\} \right\}$

The momentum vector/operator will be defined as.

$$\mathbf{p}^{\mu} = \frac{1}{\sqrt{G}} \; (n_0, \; n_1, \; n_2, \; n_3)$$

In[3]:= $\mathbf{P} = \begin{pmatrix} \frac{n_0}{\sqrt{G}} \\ \frac{n_1}{\sqrt{G}} \\ \frac{n_2}{\sqrt{G}} \\ \frac{n_3}{\sqrt{G}} \end{pmatrix}$

Out[3]= $\left\{ \left\{ \frac{n_0}{\sqrt{G}} \right\}, \; \left\{ \frac{n_1}{\sqrt{G}} \right\}, \; \left\{ \frac{n_2}{\sqrt{G}} \right\}, \; \left\{ \frac{n_3}{\sqrt{G}} \right\} \right\}$

To fundamentally define dynamics in any theory one needs to define the action. From the action, by well established means, the dynamics can be mathematically derived. In a quantum theory the important operator for dynamics is the Hamiltonian operator. In the General theory of Relativity the this role is played by the Hilbert stress energy tensor.

$$T^{\mu\nu} = \frac{1}{\sqrt{-|\chi|}} \frac{\delta \, S_{non-grav}}{\delta \, \hat{\chi}} \tag{0.19}$$

In this theory the whole stress energy tensor will play the role of the Hamiltonian operator. The stress tensor used should be basically the same form as the Hilbert stress tensor cited above. However I will change the functional derivatives to quantized space-time derivatives denoted by capital delta like so. The Hilbert stress tensor is most appropriate for this theory. It should have the same basic features as the metric tensor $\chi_{\mu\nu}$ because the metric tensor is part of the definition of the Hilbert Stress tensor.

For the purposes of this theory and in the quantized space-time described in this paper the derivatives δ, and d will be replaced with quantum derivatives Δ. These are nearly but not exactly equivalent to the traditional derivatives. Only in the finest calculations would the difference between a derivative in quantized space and smooth space be important.

$$T^{\mu\nu} = \frac{1}{\sqrt{-|\chi|}} \frac{\Delta \, S_{non-grav}}{\Delta \, \hat{\chi}} \tag{0.20}$$

Out[4]= $\{\{-1, 0, 0, 0\}, \{0, 1, 0, 0\}, \{0, 0, 1, 0\}, \{0, 0, 0, 1\}\}$

I will define the non gravitational invariant interval. Which assumes Minkowski curvature free space-time. For the purposes of this paper I will assume the simplest form for a non gravitational action $S_{non-grav}$.

In[5]:= $S_{non-grav} = \sqrt{Transpose[X] . \hat{\chi} . X \frac{1}{G}}$

In[5]:= $S_{non-grav} = \sqrt{\frac{-G \, n_0^3 + G \, n_1^3 + G \, n_2^3 + G \, n_3^3}{G}}$

Out[5]= $\sqrt{\frac{-G \, n_0^3 + G \, n_1^3 + G \, n_2^3 + G \, n_3^3}{G}}$

The functional derivatives in the definition of the Hilbert stress tensor boil down to a division operation in the special case of the action $S_{non-grav}$ I have chosen to use.

In[6]:= $\hat{T} = \frac{1}{\sqrt{-Det[\hat{\chi}] \, G}} \begin{pmatrix} -\frac{S_{non-grav}}{n_0} & 0 & 0 & 0 \\ 0 & \frac{S_{non-grav}}{n_1} & 0 & 0 \\ 0 & 0 & \frac{S_{non-grav}}{n_2} & 0 \\ 0 & 0 & 0 & \frac{S_{non-grav}}{n_3} \end{pmatrix}$

Out[6]= $\left\{ \left\{ -\frac{\sqrt{\frac{-G \, n_0^3 + G \, n_1^3 + G \, n_2^3 + G \, n_3^3}{G}}}{n_0 \sqrt{G \, n_0 \, n_1 \, n_2 \, n_3}}, 0, 0, 0 \right\}, \left\{ 0, \frac{\sqrt{\frac{-G \, n_0^3 + G \, n_1^3 + G \, n_2^3 + G \, n_3^3}{G}}}{n_1 \sqrt{G \, n_0 \, n_1 \, n_2 \, n_3}}, 0, 0 \right\}, \right.$

$\left\{ 0, 0, \frac{\sqrt{\frac{-G \, n_0^3 + G \, n_1^3 + G \, n_2^3 + G \, n_3^3}{G}}}{n_2 \sqrt{G \, n_0 \, n_1 \, n_2 \, n_3}}, 0 \right\}, \left\{ 0, 0, 0, \frac{\sqrt{\frac{-G \, n_0^3 + G \, n_1^3 + G \, n_2^3 + G \, n_3^3}{G}}}{n_3 \sqrt{G \, n_0 \, n_1 \, n_2 \, n_3}} \right\} \right\}$

The graviton operator $@_{\mu\kappa}$ ("at").

In[7]:= $\hat{at} = \hat{\chi} \, G$

Out[7]= $\{\{-G \, n_0, 0, 0, 0\}, \{0, G \, n_1, 0, 0\}, \{0, 0, G \, n_2, 0\}, \{0, 0, 0, G \, n_3\}\}$

The distribution operator.

$$a^{\mu}{}_{\kappa} = \frac{1}{2 \mid x_{\mu} \rangle \langle x^{\kappa} \mid}$$

$$\text{In[8]:= } \quad \hat{a} = \begin{pmatrix} \frac{1}{2 n_0^2 G} & \frac{1}{2 n_0 n_1 G} & \frac{1}{2 n_0 n_2 G} & \frac{1}{2 n_0 n_3 G} \\ \frac{1}{2 n_0 n_1 G} & \frac{1}{2 n_1^2 G} & \frac{1}{2 n_1 n_2 G} & \frac{1}{2 n_1 n_3 G} \\ \frac{1}{2 n_0 n_2 G} & \frac{1}{2 n_1 n_2 G} & \frac{1}{2 n_2^2 G} & \frac{1}{2 n_2 n_3 G} \\ \frac{1}{2 n_0 n_3 G} & \frac{1}{2 n_1 n_3 G} & \frac{1}{2 n_2 n_3 G} & \frac{1}{2 n_3^2 G} \end{pmatrix}$$

$$\text{Out[8]= } \left\{ \left\{ \frac{1}{2 G n_0^2}, \frac{1}{2 G n_0 n_1}, \frac{1}{2 G n_0 n_2}, \frac{1}{2 G n_0 n_3} \right\}, \left\{ \frac{1}{2 G n_0 n_1}, \frac{1}{2 G n_1^2}, \frac{1}{2 G n_1 n_2}, \frac{1}{2 G n_1 n_3} \right\}, \right.$$
$$\left. \left\{ \frac{1}{2 G n_0 n_2}, \frac{1}{2 G n_1 n_2}, \frac{1}{2 G n_2^2}, \frac{1}{2 G n_2 n_3} \right\}, \left\{ \frac{1}{2 G n_0 n_3}, \frac{1}{2 G n_1 n_3}, \frac{1}{2 G n_2 n_3}, \frac{1}{2 G n_3^2} \right\} \right\}$$

The potential of quantized space-time. $A^{\kappa\nu} = a^{\kappa}{}_{\alpha} T^{\alpha\nu}$

$$\text{In[9]:= } \quad \hat{A} = \hat{a}.\hat{T}$$

$$\text{Out[9]= } \left\{ \left\{ -\frac{\sqrt{\frac{-G n_0^3 + G n_1^3 + G n_2^3 + G n_3^3}{G}}}{2 G n_0^3 \sqrt{G n_0 n_1 n_2 n_3}}, \frac{\sqrt{\frac{-G n_0^3 + G n_1^3 + G n_2^3 + G n_3^3}{G}}}{2 G n_0 n_1^2 \sqrt{G n_0 n_1 n_2 n_3}}, \frac{\sqrt{\frac{-G n_0^3 + G n_1^3 + G n_2^3 + G n_3^3}{G}}}{2 G n_0 n_2^2 \sqrt{G n_0 n_1 n_2 n_3}}, \frac{\sqrt{\frac{-G n_0^3 + G n_1^3 + G n_2^3 + G n_3^3}{G}}}{2 G n_0 n_3^2 \sqrt{G n_0 n_1 n_2 n_3}} \right\}, \right.$$

$$\left\{ -\frac{\sqrt{\frac{-G n_0^3 + G n_1^3 + G n_2^3 + G n_3^3}{G}}}{2 G n_0^2 n_1 \sqrt{G n_0 n_1 n_2 n_3}}, \frac{\sqrt{\frac{-G n_0^3 + G n_1^3 + G n_2^3 + G n_3^3}{G}}}{2 G n_1^3 \sqrt{G n_0 n_1 n_2 n_3}}, \frac{\sqrt{\frac{-G n_0^3 + G n_1^3 + G n_2^3 + G n_3^3}{G}}}{2 G n_1 n_2^2 \sqrt{G n_0 n_1 n_2 n_3}}, \frac{\sqrt{\frac{-G n_0^3 + G n_1^3 + G n_2^3 + G n_3^3}{G}}}{2 G n_1 n_3^2 \sqrt{G n_0 n_1 n_2 n_3}} \right\},$$

$$\left\{ -\frac{\sqrt{\frac{-G n_0^3 + G n_1^3 + G n_2^3 + G n_3^3}{G}}}{2 G n_0^2 n_2 \sqrt{G n_0 n_1 n_2 n_3}}, \frac{\sqrt{\frac{-G n_0^3 + G n_1^3 + G n_2^3 + G n_3^3}{G}}}{2 G n_1^2 n_2 \sqrt{G n_0 n_1 n_2 n_3}}, \frac{\sqrt{\frac{-G n_0^3 + G n_1^3 + G n_2^3 + G n_3^3}{G}}}{2 G n_2^3 \sqrt{G n_0 n_1 n_2 n_3}}, \frac{\sqrt{\frac{-G n_0^3 + G n_1^3 + G n_2^3 + G n_3^3}{G}}}{2 G n_2 n_3^2 \sqrt{G n_0 n_1 n_2 n_3}} \right\},$$

$$\left. \left\{ -\frac{\sqrt{\frac{-G n_0^3 + G n_1^3 + G n_2^3 + G n_3^3}{G}}}{2 G n_0^2 n_3 \sqrt{G n_0 n_1 n_2 n_3}}, \frac{\sqrt{\frac{-G n_0^3 + G n_1^3 + G n_2^3 + G n_3^3}{G}}}{2 G n_1^2 n_3 \sqrt{G n_0 n_1 n_2 n_3}}, \frac{\sqrt{\frac{-G n_0^3 + G n_1^3 + G n_2^3 + G n_3^3}{G}}}{2 G n_2^2 n_3 \sqrt{G n_0 n_1 n_2 n_3}}, \frac{\sqrt{\frac{-G n_0^3 + G n_1^3 + G n_2^3 + G n_3^3}{G}}}{2 G n_3^3 \sqrt{G n_0 n_1 n_2 n_3}} \right\} \right\}$$

$$\begin{pmatrix} -\frac{\sqrt{\frac{-G n_0^3 + G n_1^3 + G n_2^3 + G n_3^3}{G}}}{2 G n_0^3 \sqrt{G n_0 n_1 n_2 n_3}} & \frac{\sqrt{\frac{-G n_0^3 + G n_1^3 + G n_2^3 + G n_3^3}{G}}}{2 G n_0 n_1^2 \sqrt{G n_0 n_1 n_2 n_3}} & \frac{\sqrt{\frac{-G n_0^3 + G n_1^3 + G n_2^3 + G n_3^3}{G}}}{2 G n_0 n_2^2 \sqrt{G n_0 n_1 n_2 n_3}} & \frac{\sqrt{\frac{-G n_0^3 + G n_1^3 + G n_2^3 + G n_3^3}{G}}}{2 G n_0 n_3^2 \sqrt{G n_0 n_1 n_2 n_3}} \\ \frac{\sqrt{\frac{-G n_0^3 + G n_1^3 + G n_2^3 + G n_3^3}{G}}}{2 G n_0^2 n_1 \sqrt{G n_0 n_1 n_2 n_3}} & \frac{\sqrt{\frac{-G n_0^3 + G n_1^3 + G n_2^3 + G n_3^3}{G}}}{2 G n_1^3 \sqrt{G n_0 n_1 n_2 n_3}} & \frac{\sqrt{\frac{-G n_0^3 + G n_1^3 + G n_2^3 + G n_3^3}{G}}}{2 G n_1 n_2^2 \sqrt{G n_0 n_1 n_2 n_3}} & \frac{\sqrt{\frac{-G n_0^3 + G n_1^3 + G n_2^3 + G n_3^3}{G}}}{2 G n_1 n_3^2 \sqrt{G n_0 n_1 n_2 n_3}} \\ \frac{\sqrt{\frac{-G n_0^3 + G n_1^3 + G n_2^3 + G n_3^3}{G}}}{2 G n_0^2 n_2 \sqrt{G n_0 n_1 n_2 n_3}} & \frac{\sqrt{\frac{-G n_0^3 + G n_1^3 + G n_2^3 + G n_3^3}{G}}}{2 G n_1^2 n_2 \sqrt{G n_0 n_1 n_2 n_3}} & \frac{\sqrt{\frac{-G n_0^3 + G n_1^3 + G n_2^3 + G n_3^3}{G}}}{2 G n_2^3 \sqrt{G n_0 n_1 n_2 n_3}} & \frac{\sqrt{\frac{-G n_0^3 + G n_1^3 + G n_2^3 + G n_3^3}{G}}}{2 G n_2 n_3^2 \sqrt{G n_0 n_1 n_2 n_3}} \\ -\frac{\sqrt{\frac{-G n_0^3 + G n_1^3 + G n_2^3 + G n_3^3}{G}}}{2 G n_0^2 n_3 \sqrt{G n_0 n_1 n_2 n_3}} & \frac{\sqrt{\frac{-G n_0^3 + G n_1^3 + G n_2^3 + G n_3^3}{G}}}{2 G n_1^2 n_3 \sqrt{G n_0 n_1 n_2 n_3}} & \frac{\sqrt{\frac{-G n_0^3 + G n_1^3 + G n_2^3 + G n_3^3}{G}}}{2 G n_2^2 n_3 \sqrt{G n_0 n_1 n_2 n_3}} & \frac{\sqrt{\frac{-G n_0^3 + G n_1^3 + G n_2^3 + G n_3^3}{G}}}{2 G n_3^3 \sqrt{G n_0 n_1 n_2 n_3}} \end{pmatrix}$$

Lastly the mass operator. $M^{\mu\nu}=@^\mu{}_\kappa T^{\kappa\nu}$ Which is also the mathematical encapsulation of the postulate of quantized space-time. It should be read as: Mass is the product of the curvature of space-time with the distribution of energy through space-time.

In[10]:= $\hat{\mathbf{M}} = \hat{\mathbf{at}} . \hat{\mathbf{A}}$

Out[10]= $\left\{\left\{\dfrac{\sqrt{\dfrac{-G\,n_0^3+G\,n_1^3+G\,n_2^3+G\,n_3^3}{G}}}{2\,n_0^2\,\sqrt{G\,n_0\,n_1\,n_2\,n_3}},\; -\dfrac{\sqrt{\dfrac{-G\,n_0^3+G\,n_1^3+G\,n_2^3+G\,n_3^3}{G}}}{2\,n_1^2\,\sqrt{G\,n_0\,n_1\,n_2\,n_3}},\; -\dfrac{\sqrt{\dfrac{-G\,n_0^3+G\,n_1^3+G\,n_2^3+G\,n_3^3}{G}}}{2\,n_2^2\,\sqrt{G\,n_0\,n_1\,n_2\,n_3}},\; -\dfrac{\sqrt{\dfrac{-G\,n_0^3+G\,n_1^3+G\,n_2^3+G\,n_3^3}{G}}}{2\,n_3^2\,\sqrt{G\,n_0\,n_1\,n_2\,n_3}}\right\},\right.$

$\left\{-\dfrac{\sqrt{\dfrac{-G\,n_0^3+G\,n_1^3+G\,n_2^3+G\,n_3^3}{G}}}{2\,n_0^2\,\sqrt{G\,n_0\,n_1\,n_2\,n_3}},\; \dfrac{\sqrt{\dfrac{-G\,n_0^3+G\,n_1^3+G\,n_2^3+G\,n_3^3}{G}}}{2\,n_1^2\,\sqrt{G\,n_0\,n_1\,n_2\,n_3}},\; \dfrac{\sqrt{\dfrac{-G\,n_0^3+G\,n_1^3+G\,n_2^3+G\,n_3^3}{G}}}{2\,n_2^2\,\sqrt{G\,n_0\,n_1\,n_2\,n_3}},\; \dfrac{\sqrt{\dfrac{-G\,n_0^3+G\,n_1^3+G\,n_2^3+G\,n_3^3}{G}}}{2\,n_3^2\,\sqrt{G\,n_0\,n_1\,n_2\,n_3}}\right\},$

$\left\{-\dfrac{\sqrt{\dfrac{-G\,n_0^3+G\,n_1^3+G\,n_2^3+G\,n_3^3}{G}}}{2\,n_0^2\,\sqrt{G\,n_0\,n_1\,n_2\,n_3}},\; \dfrac{\sqrt{\dfrac{-G\,n_0^3+G\,n_1^3+G\,n_2^3+G\,n_3^3}{G}}}{2\,n_1^2\,\sqrt{G\,n_0\,n_1\,n_2\,n_3}},\; \dfrac{\sqrt{\dfrac{-G\,n_0^3+G\,n_1^3+G\,n_2^3+G\,n_3^3}{G}}}{2\,n_2^2\,\sqrt{G\,n_0\,n_1\,n_2\,n_3}},\; \dfrac{\sqrt{\dfrac{-G\,n_0^3+G\,n_1^3+G\,n_2^3+G\,n_3^3}{G}}}{2\,n_3^2\,\sqrt{G\,n_0\,n_1\,n_2\,n_3}}\right\},$

$\left.\left\{-\dfrac{\sqrt{\dfrac{-G\,n_0^3+G\,n_1^3+G\,n_2^3+G\,n_3^3}{G}}}{2\,n_0^2\,\sqrt{G\,n_0\,n_1\,n_2\,n_3}},\; \dfrac{\sqrt{\dfrac{-G\,n_0^3+G\,n_1^3+G\,n_2^3+G\,n_3^3}{G}}}{2\,n_1^2\,\sqrt{G\,n_0\,n_1\,n_2\,n_3}},\; \dfrac{\sqrt{\dfrac{-G\,n_0^3+G\,n_1^3+G\,n_2^3+G\,n_3^3}{G}}}{2\,n_2^2\,\sqrt{G\,n_0\,n_1\,n_2\,n_3}},\; \dfrac{\sqrt{\dfrac{-G\,n_0^3+G\,n_1^3+G\,n_2^3+G\,n_3^3}{G}}}{2\,n_3^2\,\sqrt{G\,n_0\,n_1\,n_2\,n_3}}\right\}\right\}$

It is from this equation that all the gravitational physics flows.

Operator Algebra.

With it proven that this space is a Hilbert space which is differentiable I can now define the algebra of the operators in terms of Poission brackets. Note the traces of these commutators. Even though the full operator commutators of these quantities do not commute the scalars do. So there is no contradiction between this theory and observed experience with quantities we have all assumed are best characterized by scalars.

Now to have the computer work out the commutators. In all cases I will define what the quantities in the commutator are. I will assign symbols to those quantities. Those symbols will be used for algebraic purposes. In the cases where the matrix form output will not fit on a page I will shrink it. The really important thing to remember for using operator algebra is what commutes and what does not commute. The exact form of the non commuting result is unimportant.

I will begin with the commutator which is fundamental to dynamics in the quantized space-time that I have defined in this paper.

[@, T]

In[27]:= $\hat{\mathbf{at}} . \hat{\mathbf{T}} - \hat{\mathbf{T}} . \hat{\mathbf{at}}$

Out[27]= $\{\{0, 0, 0, 0\}, \{0, 0, 0, 0\}, \{0, 0, 0, 0\}, \{0, 0, 0, 0\}\}$

Looking at the dimensions of this quantity it has the dimensions of length. What should I call this? I have written a whole section about this operator so I will not elaborate on this at this time. See the section on dynamics.

[@, A]

@A has been defined above as mass. This is part of the postulate of quantum space-time. There does not seem to be any simple relationship between $\hat{\mathbf{at}}.\hat{\mathbf{A}}$ and $\hat{\mathbf{A}}.\hat{\mathbf{at}}$ I will call $\hat{\mathbf{A}}.\hat{\mathbf{at}} = \hat{\lambda}$

In[28]:= **FullSimplify$\left[\hat{at}.\hat{A} - \hat{A}.\hat{at}\right]$ // MatrixForm**

Out[28]//MatrixForm=

$$\begin{pmatrix} 0 & -\dfrac{(n_0+n_1)\sqrt{-n_0^3+n_1^3+n_2^3+n_3^3}}{2\,n_0\,n_1^2\sqrt{G\,n_0\,n_1\,n_2\,n_3}} & -\dfrac{(n_0+n_2)\sqrt{-n_0^3+n_1^3+n_2^3+n_3^3}}{2\,n_0\,n_2^2\sqrt{G\,n_0\,n_1\,n_2\,n_3}} & -\dfrac{(n_0+n_3)\sqrt{-n_0^3+n_1^3+n_2^3+n_3^3}}{2\,n_0\,n_3^2\sqrt{G\,n_0\,n_1\,n_2\,n_3}} \\[2.2em] -\dfrac{(n_0+n_1)\sqrt{-n_0^3+n_1^3+n_2^3+n_3^3}}{2\,n_0^2\,n_1\sqrt{G\,n_0\,n_1\,n_2\,n_3}} & 0 & \dfrac{(n_1-n_2)\sqrt{-n_0^3+n_1^3+n_2^3+n_3^3}}{2\,n_1\,n_2^2\sqrt{G\,n_0\,n_1\,n_2\,n_3}} & \dfrac{(n_1-n_3)\sqrt{-n_0^3+n_1^3+n_2^3+n_3^3}}{2\,n_1\,n_3^2\sqrt{G\,n_0\,n_1\,n_2\,n_3}} \\[2.2em] -\dfrac{(n_0+n_2)\sqrt{-n_0^3+n_1^3+n_2^3+n_3^3}}{2\,n_0^2\,n_2\sqrt{G\,n_0\,n_1\,n_2\,n_3}} & \dfrac{(-n_1+n_2)\sqrt{-n_0^3+n_1^3+n_2^3+n_3^3}}{2\,n_1^2\,n_2\sqrt{G\,n_0\,n_1\,n_2\,n_3}} & 0 & \dfrac{(n_2-n_3)\sqrt{-n_0^3+n_1^3+n_2^3+n_3^3}}{2\,n_2\,n_3^2\sqrt{G\,n_0\,n_1\,n_2\,n_3}} \\[2.2em] -\dfrac{(n_0+n_3)\sqrt{-n_0^3+n_1^3+n_2^3+n_3^3}}{2\,n_0^2\,n_3\sqrt{G\,n_0\,n_1\,n_2\,n_3}} & \dfrac{(-n_1+n_3)\sqrt{-n_0^3+n_1^3+n_2^3+n_3^3}}{2\,n_1^2\,n_3\sqrt{G\,n_0\,n_1\,n_2\,n_3}} & \dfrac{(-n_2+n_3)\sqrt{-n_0^3+n_1^3+n_2^3+n_3^3}}{2\,n_2^2\,n_3\sqrt{G\,n_0\,n_1\,n_2\,n_3}} & 0 \end{pmatrix}$$

In[115]:= **[@, M]**

In[29]:= **FullSimplify$\left[\hat{at}.\hat{M} - \hat{M}.\hat{at}\right]$ // MatrixForm**

Out[29]//MatrixForm=

$$\begin{pmatrix} 0 & \dfrac{G\,(n_0+n_1)\sqrt{-n_0^3+n_1^3+n_2^3+n_3^3}}{2\,n_1^2\sqrt{G\,n_0\,n_1\,n_2\,n_3}} & \dfrac{G\,(n_0+n_2)\sqrt{-n_0^3+n_1^3+n_2^3+n_3^3}}{2\,n_2^2\sqrt{G\,n_0\,n_1\,n_2\,n_3}} & \dfrac{G\,(n_0+n_3)\sqrt{-n_0^3+n_1^3+n_2^3+n_3^3}}{2\,n_3^2\sqrt{G\,n_0\,n_1\,n_2\,n_3}} \\[2.2em] -\dfrac{G\,(n_0+n_1)\sqrt{-n_0^3+n_1^3+n_2^3+n_3^3}}{2\,n_0^2\sqrt{G\,n_0\,n_1\,n_2\,n_3}} & 0 & \dfrac{G\,(n_1-n_2)\sqrt{-n_0^3+n_1^3+n_2^3+n_3^3}}{2\,n_2^2\sqrt{G\,n_0\,n_1\,n_2\,n_3}} & \dfrac{G\,(n_1-n_3)\sqrt{-n_0^3+n_1^3+n_2^3+n_3^3}}{2\,n_3^2\sqrt{G\,n_0\,n_1\,n_2\,n_3}} \\[2.2em] -\dfrac{G\,(n_0+n_2)\sqrt{-n_0^3+n_1^3+n_2^3+n_3^3}}{2\,n_0^2\sqrt{G\,n_0\,n_1\,n_2\,n_3}} & \dfrac{G\,(-n_1+n_2)\sqrt{-n_0^3+n_1^3+n_2^3+n_3^3}}{2\,n_1^2\sqrt{G\,n_0\,n_1\,n_2\,n_3}} & 0 & \dfrac{G\,(n_2-n_3)\sqrt{-n_0^3+n_1^3+n_2^3+n_3^3}}{2\,n_3^2\sqrt{G\,n_0\,n_1\,n_2\,n_3}} \\[2.2em] -\dfrac{G\,(n_0+n_3)\sqrt{-n_0^3+n_1^3+n_2^3+n_3^3}}{2\,n_0^2\sqrt{G\,n_0\,n_1\,n_2\,n_3}} & \dfrac{G\,(-n_1+n_3)\sqrt{-n_0^3+n_1^3+n_2^3+n_3^3}}{2\,n_1^2\sqrt{G\,n_0\,n_1\,n_2\,n_3}} & \dfrac{G\,(-n_2+n_3)\sqrt{-n_0^3+n_1^3+n_2^3+n_3^3}}{2\,n_2^2\sqrt{G\,n_0\,n_1\,n_2\,n_3}} & 0 \end{pmatrix}$$

In[30]:= **Tr$\left[\hat{at}.\hat{M} - \hat{M}.\hat{at}\right]$**

Out[30]= 0

[@, x]

In[31]:= **$\hat{at}.X$ - Transpose$\left[\text{Transpose}[X].\hat{at}\right]$**

Out[31]= {{0}, {0}, {0}, {0}}

[@, p]

In[32]:= **$\hat{at}.p$ - Transpose$\left[\text{Transpose}[p].\hat{at}\right]$**

Out[32]= {{0}, {0}, {0}, {0}}

[A, T]

In[33]:= **FullSimplify$\left[\hat{A}.\hat{T} - \hat{T}.\hat{A}\right]$ // MatrixForm**

Out[33]//MatrixForm=

$$
\begin{pmatrix}
0 & -\dfrac{(n_0+n_1)\left(n_0^3-n_1^3-n_2^3-n_3^3\right)}{2\,G^2\,n_0^3\,n_1^4\,n_2\,n_3} & -\dfrac{(n_0+n_2)\left(n_0^3-n_1^3-n_2^3-n_3^3\right)}{2\,G^2\,n_0^3\,n_1\,n_2^4\,n_3} & -\dfrac{(n_0+n_3)\left(n_0^3-n_1^3-n_2^3-n_3^3\right)}{2\,G^2\,n_0^3\,n_1\,n_2\,n_3^4} \\[3ex]
-\dfrac{(n_0+n_1)\left(n_0^3-n_1^3-n_2^3-n_3^3\right)}{2\,G^2\,n_0^4\,n_1^3\,n_2\,n_3} & 0 & \dfrac{(n_1-n_2)\left(-n_0^3+n_1^3+n_2^3+n_3^3\right)}{2\,G^2\,n_0\,n_1^3\,n_2^4\,n_3} & \dfrac{(n_1-n_3)\left(-n_0^3+n_1^3+n_2^3+n_3^3\right)}{2\,G^2\,n_0\,n_1^3\,n_2\,n_3^4} \\[3ex]
-\dfrac{(n_0+n_2)\left(n_0^3-n_1^3-n_2^3-n_3^3\right)}{2\,G^2\,n_0^4\,n_1\,n_2^3\,n_3} & -\dfrac{(n_1-n_2)\left(-n_0^3+n_1^3+n_2^3+n_3^3\right)}{2\,G^2\,n_0\,n_1^4\,n_2^3\,n_3} & 0 & \dfrac{(n_2-n_3)\left(-n_0^3+n_1^3+n_2^3+n_3^3\right)}{2\,G^2\,n_0\,n_1\,n_2^3\,n_3^4} \\[3ex]
-\dfrac{(n_0+n_3)\left(n_0^3-n_1^3-n_2^3-n_3^3\right)}{2\,G^2\,n_0^4\,n_1\,n_2\,n_3^3} & -\dfrac{(n_1-n_3)\left(-n_0^3+n_1^3+n_2^3+n_3^3\right)}{2\,G^2\,n_0\,n_1^4\,n_2\,n_3^3} & -\dfrac{(n_2-n_3)\left(-n_0^3+n_1^3+n_2^3+n_3^3\right)}{2\,G^2\,n_0\,n_1\,n_2^4\,n_3^3} & 0
\end{pmatrix}
$$

In[34]:= **Tr$\left[\hat{A}.\hat{T} - \hat{T}.\hat{A}\right]$**

Out[34]= 0

[A, x]

In[35]:= **FullSimplify$\left[\hat{A}.X - \text{Transpose}\left[\text{Transpose}[X].\hat{A}\right]\right]$ // MatrixForm**

Out[35]//MatrixForm=

$$
\begin{pmatrix}
\dfrac{\sqrt{G}\,\sqrt{-n_0^3+n_1^3+n_2^3+n_3^3}\,\left(3\,n_1\,n_2\,n_3+n_0\,(n_2\,n_3+n_1\,(n_2+n_3))\right)}{2\,n_0\,(G\,n_0\,n_1\,n_2\,n_3)^{3/2}} \\[3ex]
\dfrac{\sqrt{G}\,\sqrt{-n_0^3+n_1^3+n_2^3+n_3^3}\,\left(n_1\,n_2\,n_3+n_0\,(-n_2\,n_3+n_1\,(n_2+n_3))\right)}{2\,n_1\,(G\,n_0\,n_1\,n_2\,n_3)^{3/2}} \\[3ex]
\dfrac{\sqrt{G}\,\sqrt{-n_0^3+n_1^3+n_2^3+n_3^3}\,\left(n_1\,n_2\,n_3+n_0\,(n_1\,(n_2-n_3)+n_2\,n_3)\right)}{2\,n_2\,(G\,n_0\,n_1\,n_2\,n_3)^{3/2}} \\[3ex]
\dfrac{\sqrt{G}\,\sqrt{-n_0^3+n_1^3+n_2^3+n_3^3}\,\left(n_1\,n_2\,n_3+n_0\,(-n_1\,n_2+(n_1+n_2)\,n_3)\right)}{2\,n_3\,(G\,n_0\,n_1\,n_2\,n_3)^{3/2}}
\end{pmatrix}
$$

[A, p]

In[36]:= **FullSimplify$\left[\hat{A}.p - \text{Transpose}\left[\text{Transpose}[p].\hat{A}\right]\right]$ // MatrixForm**

Out[36]//MatrixForm=

$$
\begin{pmatrix}
\dfrac{\sqrt{-n_0^3+n_1^3+n_2^3+n_3^3}\,\left(3\,n_1\,n_2\,n_3+n_0\,(n_2\,n_3+n_1\,(n_2+n_3))\right)}{2\,\sqrt{G}\,n_0\,(G\,n_0\,n_1\,n_2\,n_3)^{3/2}} \\[3ex]
\dfrac{\sqrt{-n_0^3+n_1^3+n_2^3+n_3^3}\,\left(-n_1\,n_2\,n_3+n_0\,(-3\,n_2\,n_3+n_1\,(n_2+n_3))\right)}{2\,\sqrt{G}\,n_1\,(G\,n_0\,n_1\,n_2\,n_3)^{3/2}} \\[3ex]
\dfrac{\sqrt{-n_0^3+n_1^3+n_2^3+n_3^3}\,\left(-n_1\,n_2\,n_3+n_0\,(n_1\,(n_2-3\,n_3)+n_2\,n_3)\right)}{2\,\sqrt{G}\,n_2\,(G\,n_0\,n_1\,n_2\,n_3)^{3/2}} \\[3ex]
\dfrac{\sqrt{-n_0^3+n_1^3+n_2^3+n_3^3}\,\left(-n_1\,n_2\,n_3+n_0\,(-3\,n_1\,n_2+(n_1+n_2)\,n_3)\right)}{2\,\sqrt{G}\,n_3\,(G\,n_0\,n_1\,n_2\,n_3)^{3/2}}
\end{pmatrix}
$$

[A, M]

In[37]:= **FullSimplify$\left[\hat{A}.\hat{M} - \hat{M}.\hat{A}\right]$ // MatrixForm**

Out[37]//MatrixForm=

$$
\begin{pmatrix}
\frac{\left(n_0^3-n_1^3-n_2^3-n_3^3\right)\left(n_0\,n_1^3\,n_2^3+n_1^3\,n_2^3\,n_3+\left(n_0\,n_1^3+n_1^3\,n_2+(n_0+n_1)\,n_2^3\right)n_3^3\right)}{4\,G^2\,n_0^4\,n_1^4\,n_2^4\,n_3^4} & -\frac{\left(n_0^3-n_1^3-n_2^3-n_3^3\right)\left(n_0\,n_1^3\,n_2^3+n_1^3\,n_2^3\,n_3+\left(n_0\,n_1^3+n_1^3\,n_2+(n_0+n_1)\,n_2^3\right)n_3^3\right)}{4\,G^2\,n_0^6\,n_1^4\,n_2^4\,n_3^4} & -\frac{\left(n_0^3-n_1^3-n_2^3-n\right)}{} \\
-\frac{\left(n_0^3-n_1^3-n_2^3-n_3^3\right)\left(n_0^3\,n_1\,n_2^3-n_0^3\,n_2^3\,n_3+\left(n_0^3\,n_1-n_0^3\,n_2-(n_0+n_1)\,n_2^3\right)n_3^3\right)}{4\,G^2\,n_0^6\,n_1^2\,n_2^4\,n_3^4} & \frac{\left(n_0^3-n_1^3-n_2^3-n_3^3\right)\left(n_0^3\,n_1\,n_2^3-n_0^3\,n_2^3\,n_3+\left(n_0^3\,n_1-n_0^3\,n_2-(n_0+n_1)\,n_2^3\right)n_3^3\right)}{4\,G^2\,n_0^4\,n_1^2\,n_2^4\,n_3^4} & \frac{\left(n_0^3-n_1^3-n_2^3-n_3^3\right)}{} \\
-\frac{\left(n_0^3-n_1^3-n_2^3-n_3^3\right)\left(n_0^3\,n_1^3\,n_2-n_0^3\,n_1^3\,n_3-\left(n_0\,n_1^3+n_0^3\,(n_1-n_2)+n_1^3\,n_2\right)n_3^3\right)}{4\,G^2\,n_0^6\,n_1^4\,n_2^2\,n_3^4} & \frac{\left(n_0^3-n_1^3-n_2^3-n_3^3\right)\left(n_0^3\,n_1^3\,n_2-n_0^3\,n_1^3\,n_3-\left(n_0\,n_1^3+n_0^3\,(n_1-n_2)+n_1^3\,n_2\right)n_3^3\right)}{4\,G^2\,n_0^4\,n_1^6\,n_2^2\,n_3^4} & \frac{\left(n_0^3-n_1^3-n_2^3-n_3^3\right)}{} \\
\frac{\left(n_0^3-n_1^3-n_2^3-n_3^3\right)\left(n_0\,n_1^3\,n_2^3+n_1^3\,n_2^3\,n_3+n_0^3\,\left(n_1\,n_2\,\left(n_1^2+n_2^2\right)-\left(n_1^3+n_2^3\right)n_3\right)\right)}{4\,G^2\,n_0^6\,n_1^4\,n_2^4\,n_3^2} & -\frac{\left(n_0^3-n_1^3-n_2^3-n_3^3\right)\left(n_0\,n_1^3\,n_2^3+n_1^3\,n_2^3\,n_3+n_0^3\,\left(n_1\,n_2\,\left(n_1^2+n_2^2\right)-\left(n_1^3+n_2^3\right)n_3\right)\right)}{4\,G^2\,n_0^4\,n_1^4\,n_2^4\,n_3^2} & -\frac{\left(n_0^3-n_1^3-n_2^3-n\right)}{}
\end{pmatrix}
$$

In[38]:= **Tr$\left[\hat{A}.\hat{M} - \hat{M}.\hat{A}\right]$**

Out[38]= 0

[M, T]

In[39]:= **FullSimplify$\left[\hat{M}.\hat{T} - \hat{T}.\hat{M}\right]$ // MatrixForm**

Out[39]//MatrixForm=

$$
\begin{pmatrix}
0 & \frac{(n_0+n_1)\left(n_0^3-n_1^3-n_2^3-n_3^3\right)}{2\,G\,n_0^2\,n_1^4\,n_2\,n_3} & \frac{(n_0+n_2)\left(n_0^3-n_1^3-n_2^3-n_3^3\right)}{2\,G\,n_0^2\,n_1\,n_2^4\,n_3} & \frac{(n_0+n_3)\left(n_0^3-n_1^3-n_2^3-n_3^3\right)}{2\,G\,n_0^2\,n_1\,n_2\,n_3^4} \\
-\frac{(n_0+n_1)\left(n_0^3-n_1^3-n_2^3-n_3^3\right)}{2\,G\,n_0^4\,n_1^2\,n_2\,n_3} & 0 & \frac{(n_1-n_2)\left(-n_0^3+n_1^3+n_2^3+n_3^3\right)}{2\,G\,n_0\,n_1^2\,n_2^4\,n_3} & \frac{(n_1-n_3)\left(-n_0^3+n_1^3+n_2^3+n_3^3\right)}{2\,G\,n_0\,n_1^2\,n_2\,n_3^4} \\
-\frac{(n_0+n_2)\left(n_0^3-n_1^3-n_2^3-n_3^3\right)}{2\,G\,n_0^4\,n_1\,n_2^2\,n_3} & -\frac{(n_1-n_2)\left(-n_0^3+n_1^3+n_2^3+n_3^3\right)}{2\,G\,n_0\,n_1^4\,n_2^2\,n_3} & 0 & \frac{(n_2-n_3)\left(-n_0^3+n_1^3+n_2^3+n_3^3\right)}{2\,G\,n_0\,n_1\,n_2^2\,n_3^4} \\
-\frac{(n_0+n_3)\left(n_0^3-n_1^3-n_2^3-n_3^3\right)}{2\,G\,n_0^4\,n_1\,n_2\,n_3^2} & -\frac{(n_1-n_3)\left(-n_0^3+n_1^3+n_2^3+n_3^3\right)}{2\,G\,n_0\,n_1^4\,n_2\,n_3^2} & -\frac{(n_2-n_3)\left(-n_0^3+n_1^3+n_2^3+n_3^3\right)}{2\,G\,n_0\,n_1\,n_2^4\,n_3^2} & 0
\end{pmatrix}
$$

In[40]:= **Tr$\left[\hat{M}.\hat{T} - \hat{T}.\hat{M}\right]$**

Out[40]= 0

[M, x]

In[41]:= **FullSimplify$\left[\hat{M}.X - \text{Transpose}\left[\text{Transpose}[X].\hat{M}\right]\right]$ // MatrixForm**

Out[41]//MatrixForm=

$$
\begin{pmatrix}
\frac{G^{3/2}\sqrt{-n_0^3+n_1^3+n_2^3+n_3^3}\,\left(n_1\,n_2\,n_3\,(n_1+n_2+n_3)-n_0^2\,(n_2\,n_3+n_1\,(n_2+n_3))\right)}{2\,n_0\,(G\,n_0\,n_1\,n_2\,n_3)^{3/2}} \\
\frac{G^{3/2}\sqrt{-n_0^3+n_1^3+n_2^3+n_3^3}\,\left(-n_0^2\,n_2\,n_3+n_1^2\,n_2\,n_3+n_0\,(n_2+n_3)\left(n_1^2-n_2\,n_3\right)\right)}{2\,n_1\,(G\,n_0\,n_1\,n_2\,n_3)^{3/2}} \\
\frac{G^{3/2}\sqrt{-n_0^3+n_1^3+n_2^3+n_3^3}\,\left(-n_0^2\,n_1\,n_3+n_1\,n_2^2\,n_3-n_0\,(n_1+n_3)\left(-n_2^2+n_1\,n_3\right)\right)}{2\,n_2\,(G\,n_0\,n_1\,n_2\,n_3)^{3/2}} \\
\frac{G^{3/2}\left(-n_0\,n_1\,n_2\,(n_0+n_1+n_2)+(n_1\,n_2+n_0\,(n_1+n_2))\,n_3^2\right)\sqrt{-n_0^3+n_1^3+n_2^3+n_3^3}}{2\,n_3\,(G\,n_0\,n_1\,n_2\,n_3)^{3/2}}
\end{pmatrix}
$$

In[67]:= **[M, p]**

In[42]:= **FullSimplify$\left[\hat{A}.p - Transpose\left[Transpose[p].\hat{A}\right]\right]$ // MatrixForm**

Out[42]//MatrixForm=

$$
\begin{pmatrix}
\dfrac{\sqrt{-n_0^3+n_1^3+n_2^3+n_3^3}\ (3\,n_1\,n_2\,n_3+n_0\,(n_2\,n_3+n_1\,(n_2+n_3)))}{2\sqrt{G}\ n_0\,(G\,n_0\,n_1\,n_2\,n_3)^{3/2}} \\[4ex]
\dfrac{\sqrt{-n_0^3+n_1^3+n_2^3+n_3^3}\ (-n_1\,n_2\,n_3+n_0\,(-3\,n_2\,n_3+n_1\,(n_2+n_3)))}{2\sqrt{G}\ n_1\,(G\,n_0\,n_1\,n_2\,n_3)^{3/2}} \\[4ex]
\dfrac{\sqrt{-n_0^3+n_1^3+n_2^3+n_3^3}\ (-n_1\,n_2\,n_3+n_0\,(n_1\,(n_2-3\,n_3)+n_2\,n_3))}{2\sqrt{G}\ n_2\,(G\,n_0\,n_1\,n_2\,n_3)^{3/2}} \\[4ex]
\dfrac{\sqrt{-n_0^3+n_1^3+n_2^3+n_3^3}\ (-n_1\,n_2\,n_3+n_0\,(-3\,n_1\,n_2+(n_1+n_2)\,n_3))}{2\sqrt{G}\ n_3\,(G\,n_0\,n_1\,n_2\,n_3)^{3/2}}
\end{pmatrix}
$$

Now that it is clear what commutes with what algebra can be employed in problems dealing with this theory.

Mechanics in Quantized Space-Time.

Consider the commutator of @ and T. Note the dimension of the product of these operators. It has the dimension of length. Define @T=S. Then note that $@T^{\dagger} = @T$. Where the $\dagger$ indicates the conjugate transpose operation. Therefore by the accepted structure of quantum physical theories S must be an observable quantity. That this is so can be easily deduced from examination of the above presented commutators or by running this *Mathematica* notebook for yourself.

$$[@, T] = @T - T@ = S - S^{\dagger} = 0 \tag{0.21}$$

There is only one quantity in gravitational theory left for S to represent with the dimension of length. Therefore S shall be known as the gravitational path length operator. This operator has the units of length and it's eigenvalues are the possible path lengths in the theory of quantum space-time. Which are the soul of a relativistic formulation of mechanics, and quantum mechanics. What solving for the action operator will give is basically a set of eigen paths. Which is what I will call the eigenstates of S. So by solving for the eigenvalues and eigenvectors of S for a system we are finding the path's that it is possible for the system to follow. As with other observable operators we can compute the expectation values. Due to the simple action I have assumed the form of the S operator is very simple.

In[45]:= $\hat{at}.\hat{T}$ // **MatrixForm**

Out[45]//MatrixForm=

$$\begin{pmatrix}
G\sqrt{\dfrac{-G n_0^3 + G n_1^3 + G n_2^3 + G n_3^3}{G}} \Big/ \sqrt{G n_0 n_1 n_2 n_3} & 0 & 0 & 0 \\
0 & G\sqrt{\dfrac{-G n_0^3 + G n_1^3 + G n_2^3 + G n_3^3}{G}} \Big/ \sqrt{G n_0 n_1 n_2 n_3} & 0 & 0 \\
0 & 0 & G\sqrt{\dfrac{G n_0^3 + G n_1^3 + G n_2^3 + G n_3^3}{G}} \Big/ \sqrt{G n_0 n_1 n_2 n_3} & 0 \\
0 & 0 & 0 & G\sqrt{\dfrac{-G n_0^3 + G n_1^3 + G n_2^3 + G n_3^3}{G}} \Big/ \sqrt{G n_0 n_1 n_2 n_3}
\end{pmatrix}$$

The eigenvalues and eigenvectors are trivially obvious.

In[46]:= **Eigensystem** $\left[\hat{at}.\hat{T}\right]$

Out[46]= $\left\{\left\{\dfrac{G\sqrt{-n_0^3 + n_1^3 + n_2^3 + n_3^3}}{\sqrt{G n_0 n_1 n_2 n_3}}, \dfrac{G\sqrt{-n_0^3 + n_1^3 + n_2^3 + n_3^3}}{\sqrt{G n_0 n_1 n_2 n_3}}, \dfrac{G\sqrt{-n_0^3 + n_1^3 + n_2^3 + n_3^3}}{\sqrt{G n_0 n_1 n_2 n_3}}, \dfrac{G\sqrt{-n_0^3 + n_1^3 + n_2^3 + n_3^3}}{\sqrt{G n_0 n_1 n_2 n_3}}\right\},\right.$

$\left.\{\{0, 0, 0, 1\}, \{0, 0, 1, 0\}, \{0, 1, 0, 0\}, \{1, 0, 0, 0\}\}\right\}$

The eigenvalues are four fold degenerate. In this situation all path's are equal. I know this composition is probably not in front of you the reader in the form of a Mathematica notebook. The beauty of a properly written notebook is that if one needs to rerun a calculation one can simply change some definitions, and hit evaluate. Notice that when all the coefficients of geometry n_i are equal to one these eigenvalues all simplify down to the invariant interval in Minkowski space-time. Thus confirming the conjecture that the S operator corresponds to the path length. Basically this operator allows us to solve for the possible metrics by the methods employed in quantum mechanics. This seems to be just what we have been looking for.

Quantum Geometric Gravity due to Planck Scale Masses.

This is an exposition of the problem solving power of the theory of Quantized space-time. I will derive an exact non-perturbative solution for the line elements in the near neighborhood of a uniform planck scale mass. The derivation will be fully quantum and fully relativistic. That it connects to gravity at all will only be clear in the last steps.

These are some questions that will certainly come to the fore of the readers mind. "Why would I pick the Planck scale for the masses? Why not any arbitrary number? Clearly masses of all conceivable sizes exist. " I have shown in The previous paper that a mass of exactly one planck mass will collapse into a black hole with a Schwarzschild radius of one Planck length. While many masses are smaller than the Planck mass no measurable length can be smaller than the Planck length. The Planck length is defined in terms of the fundamental generally invariant constants $\hbar$ c and G. So the Planck length is in fact a Pioncare invariant interval of length. What this means is that no mass smaller than the Planck mass may gravitationally collapse. **Furthermore no single unbound mass which is less than the planck mass may have a gravitational field. Furthermore only units of one planck mass or greater may have a gravitational field.** This is a restatement of the postulate of quantized space- time. Let us see where the math it implies leads us. Let that be the test of the postulates validity.

We will do this by determining the gravitational field due to a Planck scale mass which is stationary and has no spin. The simplest way to do this would be to start with a arbitrary action functional of the factors of geometry (the n's). This will give the most general form of the eigen paths due to a mass in motion or an interaction.

In[11]:= $S_{non-grav} = \alpha[n_0, n_1, n_2, n_3]$

Out[11]= $\alpha[n_0, n_1, n_2, n_3]$

The Minkowski metric is used because an assumption of this theory is that individual quantums of space-time have no curvature.

In[12]:= $g = \begin{pmatrix} -1 & 0 & 0 & 0 \\ 0 & 1 & 0 & 0 \\ 0 & 0 & 1 & 0 \\ 0 & 0 & 0 & 1 \end{pmatrix}$

Out[12]= $\{\{-1, 0, 0, 0\}, \{0, 1, 0, 0\}, \{0, 0, 1, 0\}, \{0, 0, 0, 1\}\}$

The following is an arbitrary dynamical metric. The n's are integers.

In[13]:= $\hat{\chi} = \begin{pmatrix} -n_0 & 0 & 0 & 0 \\ 0 & n_1 & 0 & 0 \\ 0 & 0 & n_2 & 0 \\ 0 & 0 & 0 & n_3 \end{pmatrix}$

Out[13]= $\{\{-n_0, 0, 0, 0\}, \{0, n_1, 0, 0\}, \{0, 0, n_2, 0\}, \{0, 0, 0, n_3\}\}$

The quantum stress energy tensor for this section of the text will be.

In[14]:= $\hat{T} = \frac{1}{\sqrt{-Det[\hat{\chi}] G}} \begin{pmatrix} -\frac{S_{non-grav}}{n_0} & 0 & 0 & 0 \\ 0 & \frac{S_{non-grav}}{n_1} & 0 & 0 \\ 0 & 0 & \frac{S_{non-grav}}{n_2} & 0 \\ 0 & 0 & 0 & \frac{S_{non-grav}}{n_3} \end{pmatrix}$

Out[14]= $\left\{\left\{-\frac{\alpha[n_0, n_1, n_2, n_3]}{n_0 \sqrt{G n_0 n_1 n_2 n_3}}, 0, 0, 0\right\}, \left\{0, \frac{\alpha[n_0, n_1, n_2, n_3]}{n_1 \sqrt{G n_0 n_1 n_2 n_3}}, 0, 0\right\},\right.$
$\left.\left\{0, 0, \frac{\alpha[n_0, n_1, n_2, n_3]}{n_2 \sqrt{G n_0 n_1 n_2 n_3}}, 0\right\}, \left\{0, 0, 0, \frac{\alpha[n_0, n_1, n_2, n_3]}{n_3 \sqrt{G n_0 n_1 n_2 n_3}}\right\}\right\}$

The graviton operator. (named so for obvious reasons)

In[15]:= $\hat{at} = g\, G$

Out[15]= $\{\{-G, 0, 0, 0\}, \{0, G, 0, 0\}, \{0, 0, G, 0\}, \{0, 0, 0, G\}\}$

Last but not least the path length operator. This will give the possible path lengths due to this arbitrary action functional α .

In[16]:= **a͡t.T̂ // MatrixForm**

Out[16]//MatrixForm=

$$
\begin{pmatrix}
\dfrac{G\,\alpha[n_0,n_1,n_2,n_3]}{n_0\,\sqrt{G\,n_0\,n_1\,n_2\,n_3}} & 0 & 0 & 0 \\[2ex]
0 & \dfrac{G\,\alpha[n_0,n_1,n_2,n_3]}{n_1\,\sqrt{G\,n_0\,n_1\,n_2\,n_3}} & 0 & 0 \\[2ex]
0 & 0 & \dfrac{G\,\alpha[n_0,n_1,n_2,n_3]}{n_2\,\sqrt{G\,n_0\,n_1\,n_2\,n_3}} & 0 \\[2ex]
0 & 0 & 0 & \dfrac{G\,\alpha[n_0,n_1,n_2,n_3]}{n_3\,\sqrt{G\,n_0\,n_1\,n_2\,n_3}}
\end{pmatrix}
$$

In this most simplified case the action functional is just a scalar for all values of the n sub i. The mass of the particle m.

In[9]:= $\alpha[n_0,\ n_1,\ n_2,\ n_3]$ **= m**

Out[9]= m

In[10]:= $\hat{T} = \dfrac{1}{\sqrt{-\mathrm{Det}[\hat{\chi}]\,G}}
\begin{pmatrix}
-\dfrac{S_{\text{non-grav}}}{n_0} & 0 & 0 & 0 \\[2ex]
0 & \dfrac{S_{\text{non-grav}}}{n_1} & 0 & 0 \\[2ex]
0 & 0 & \dfrac{S_{\text{non-grav}}}{n_2} & 0 \\[2ex]
0 & 0 & 0 & \dfrac{S_{\text{non-grav}}}{n_3}
\end{pmatrix}$

Out[10]= $\left\{\left\{-\dfrac{m}{n_0\,\sqrt{G\,n_0\,n_1\,n_2\,n_3}},\ 0,\ 0,\ 0\right\},\ \left\{0,\ \dfrac{m}{n_1\,\sqrt{G\,n_0\,n_1\,n_2\,n_3}},\ 0,\ 0\right\},\right.$

$\left.\left\{0,\ 0,\ \dfrac{m}{n_2\,\sqrt{G\,n_0\,n_1\,n_2\,n_3}},\ 0\right\},\ \left\{0,\ 0,\ 0,\ \dfrac{m}{n_3\,\sqrt{G\,n_0\,n_1\,n_2\,n_3}}\right\}\right\}$

In[11]:= **a͡t.T̂ // MatrixForm**

Out[11]//MatrixForm=

$$
\begin{pmatrix}
\dfrac{G\,m}{n_0\,\sqrt{G\,n_0\,n_1\,n_2\,n_3}} & 0 & 0 & 0 \\[2ex]
0 & \dfrac{G\,m}{n_1\,\sqrt{G\,n_0\,n_1\,n_2\,n_3}} & 0 & 0 \\[2ex]
0 & 0 & \dfrac{G\,m}{n_2\,\sqrt{G\,n_0\,n_1\,n_2\,n_3}} & 0 \\[2ex]
0 & 0 & 0 & \dfrac{G\,m}{n_3\,\sqrt{G\,n_0\,n_1\,n_2\,n_3}}
\end{pmatrix}
$$

What does this result mean? First the diagonal entries in that matrix are the eigen values of the operator. These are the possible paths that are open to an object in a region of space-time where the stress energy tensor of a point-like mass is in effect. Each of these path's is like a "metric" in general relativity. Each is a possible formula for computing the gravitational potential.

Take one of these and compare it to the classical form of the gravitational potential 1/r. As any one of those n's grows to infinity with the rest remaining small the term in the denominator that is under a square root symbol becomes much less significant so this potential becomes classical at classical distance scales. Furthermore the law of gravity that I have derived becomes weaker at cosmological distances faster than at classical distances. This practically explains why the farthest galaxy's seem to be receding from us at a accelerating rate. This is the best and most straight forward indicator that this theory is the correct theory of quantized general relativity.

The Symmetry of the F(4)/Spin(4) Lie Group and Algebra.

The exceptional lie group F(4) and it's lie algebra offered me an intriguing possibility. Associated with F(4) is a four dimensional body centered cubic lattice. In my work on quantum gravity I have gone to pains to proclaim that space-time is not a lattice of fixed points. All my studies still indicate that is true. However F(4)/Spin(4) and it's "lattice" seem to have the symmetry of quantum space-time. Using the usual techniques of quantum field theory I have investigated this symmetry. This is what I found and how I found it.

The Lie group F(4) is the isometry group of a 16 dimensional riemannian manifold. It can be constructed by adding 16 spinors to SO(9). This would be too much for modeling quantum space-time. What needs to be modded out of the group are the components of Spin(4). This is so for physical reasons as Spin(4) contains transformations that denote rotating one coordinate axis into another. Which really ammounts to a redefinition of coordinate axes which should have no physical consequences. That physics is invariant under such transformations is contained in the local Lorentz symmetry of the theory. After moding out that we are left with a 46 dimensional lie group. F(4)/Spin(4) which I will denote F(4)/Spin(4).

The following are the simplest matricies that will represent F(4)/Spin(4) they are based on it's root vectors. Using the plus-minus symbol I can write all of the possiblities (i.e. If one writes +/- 1 that is really two things -1 and +1. The same concept is used below to write all 46 matricies without haaving to write 46 matricies.)

$$\text{In[17]}:= \quad \alpha = \begin{pmatrix} \pm 1 & 0 & 0 & 0 \\ 0 & \pm 1 & 0 & 0 \\ 0 & 0 & 0 & 0 \\ 0 & 0 & 0 & 0 \end{pmatrix}$$

Out[17]= $\{\{\pm 1, 0, 0, 0\}, \{0, \pm 1, 0, 0\}, \{0, 0, 0, 0\}, \{0, 0, 0, 0\}\}$

$$\text{In[18]}:= \quad \beta = \begin{pmatrix} \pm 1 & 0 & 0 & 0 \\ 0 & 0 & 0 & 0 \\ 0 & 0 & \pm 1 & 0 \\ 0 & 0 & 0 & 0 \end{pmatrix}$$

Out[18]= $\{\{\pm 1, 0, 0, 0\}, \{0, 0, 0, 0\}, \{0, 0, \pm 1, 0\}, \{0, 0, 0, 0\}\}$

$$\text{In[19]}:= \quad \gamma = \begin{pmatrix} \pm 1 & 0 & 0 & 0 \\ 0 & 0 & 0 & 0 \\ 0 & 0 & 0 & 0 \\ 0 & 0 & 0 & \pm 1 \end{pmatrix}$$

Out[19]= $\{\{\pm 1, 0, 0, 0\}, \{0, 0, 0, 0\}, \{0, 0, 0, 0\}, \{0, 0, 0, \pm 1\}\}$

$$\text{In[20]}:= \quad \delta = \begin{pmatrix} 0 & 0 & 0 & 0 \\ 0 & \pm 1 & 0 & 0 \\ 0 & 0 & \pm 1 & 0 \\ 0 & 0 & 0 & 0 \end{pmatrix}$$

Out[20]= $\{\{0, 0, 0, 0\}, \{0, \pm 1, 0, 0\}, \{0, 0, \pm 1, 0\}, \{0, 0, 0, 0\}\}$

$$\text{In[21]}:= \quad \epsilon = \begin{pmatrix} 0 & 0 & 0 & 0 \\ 0 & \pm 1 & 0 & 0 \\ 0 & 0 & 0 & 0 \\ 0 & 0 & 0 & \pm 1 \end{pmatrix}$$

Out[21]= $\{\{0, 0, 0, 0\}, \{0, \pm 1, 0, 0\}, \{0, 0, 0, 0\}, \{0, 0, 0, \pm 1\}\}$

In[22]:= $\varsigma = \begin{pmatrix} 0 & 0 & 0 & 0 \\ 0 & 0 & 0 & 0 \\ 0 & 0 & \pm1 & 0 \\ 0 & 0 & 0 & \pm1 \end{pmatrix}$

Out[22]= {{0, 0, 0, 0}, {0, 0, 0, 0}, {0, 0, ±1, 0}, {0, 0, 0, ±1}}

In[23]:= $\eta = \begin{pmatrix} \pm1 & 0 & 0 & 0 \\ 0 & 0 & 0 & 0 \\ 0 & 0 & 0 & 0 \\ 0 & 0 & 0 & 0 \end{pmatrix}$

Out[23]= {{±1, 0, 0, 0}, {0, 0, 0, 0}, {0, 0, 0, 0}, {0, 0, 0, 0}}

In[24]:= $\theta = \begin{pmatrix} 0 & 0 & 0 & 0 \\ 0 & \pm1 & 0 & 0 \\ 0 & 0 & 0 & 0 \\ 0 & 0 & 0 & 0 \end{pmatrix}$

Out[24]= {{0, 0, 0, 0}, {0, ±1, 0, 0}, {0, 0, 0, 0}, {0, 0, 0, 0}}

In[25]:= $\kappa = \begin{pmatrix} 0 & 0 & 0 & 0 \\ 0 & 0 & 0 & 0 \\ 0 & 0 & \pm1 & 0 \\ 0 & 0 & 0 & 0 \end{pmatrix}$

Out[25]= {{0, 0, 0, 0}, {0, 0, 0, 0}, {0, 0, ±1, 0}, {0, 0, 0, 0}}

In[26]:= $\lambda = \begin{pmatrix} 0 & 0 & 0 & 0 \\ 0 & 0 & 0 & 0 \\ 0 & 0 & 0 & 0 \\ 0 & 0 & 0 & \pm1 \end{pmatrix}$

Out[26]= {{0, 0, 0, 0}, {0, 0, 0, 0}, {0, 0, 0, 0}, {0, 0, 0, ±1}}

In[27]:= $\mu = \begin{pmatrix} \pm1 & 0 & 0 & 0 \\ 0 & \pm1 & 0 & 0 \\ 0 & 0 & \pm1 & 0 \\ 0 & 0 & 0 & \pm1 \end{pmatrix}$

Out[27]= {{±1, 0, 0, 0}, {0, ±1, 0, 0}, {0, 0, ±1, 0}, {0, 0, 0, ±1}}

In[28]:= $\Theta = \begin{pmatrix} 0 & 0 & 0 & 0 \\ 0 & 0 & 0 & 0 \\ 0 & 0 & 0 & 0 \\ 0 & 0 & 0 & 0 \end{pmatrix}$

Out[28]= {{0, 0, 0, 0}, {0, 0, 0, 0}, {0, 0, 0, 0}, {0, 0, 0, 0}}

I claim that this is a representation of the real valued variant of F(4)/Spin(4) as well as a basis for the space underlying the Lie algebra of F(4)/Spin(4). The simplest way to do this is to compute the multiplication table for the group. Right off the bat I can say that the table will be symmetric due to the fact that all of the elements of this representation are diagonal matrices. For that reason I will not bother to compute them independently.

In[14]:=

α.α	α.β	α.γ	α.δ	α.ε	α.ζ	α.η	α.θ	α.κ	α.λ	α.μ	α.Θ
□	β.β	β.γ	β.δ	β.ε	β.ζ	β.η	β.θ	β.κ	β.λ	β.μ	β.Θ
□	□	γ.γ	γ.δ	γ.ε	γ.ζ	γ.η	γ.θ	γ.κ	γ.λ	γ.μ	γ.Θ
□	□	□	δ.δ	δ.ε	δ.ζ	δ.η	δ.θ	δ.κ	δ.λ	δ.μ	δ.Θ
□	□	□	□	ε.ε	ε.ζ	ε.η	ε.θ	ε.κ	ε.λ	ε.μ	ε.Θ
□	□	□	□	□	ζ.ζ	ζ.η	ζ.θ	ζ.κ	ζ.λ	ζ.μ	ζ.Θ
□	□	□	□	□	□	η.η	η.θ	η.κ	η.λ	η.μ	η.Θ
□	□	□	□	□	□	□	θ.θ	θ.κ	θ.λ	θ.μ	θ.Θ
□	□	□	□	□	□	□	□	κ.κ	κ.λ	κ.μ	κ.Θ
□	□	□	□	□	□	□	□	□	λ.λ	λ.μ	λ.Θ
□	□	□	□	□	□	□	□	□	□	μ.μ	μ.Θ
□	□	□	□	□	□	□	□	□	□	□	Θ.Θ

// **MatrixForm**

Out[14]//MatrixForm=

$$\left(\begin{array}{cc} \{\{(\pm 1)^2, 0, 0, 0\}, \{0, (\pm 1)^2, 0, 0\}, \{0, 0, 0, 0\}, \{0, 0, 0, 0\}\} & \{\{(\pm 1)^2, 0, 0, 0\}, \{0, 0, 0, 0\}, \{0, 0, 0, 0\}, \{0 \\ & \{\{(\pm 1)^2, 0, 0, 0\}, \{0, 0, 0, 0\}, \{0, 0, (\pm 1)^2, 0\}, \\ \square & \square \\ \square & \square \\ \square & \square \\ \square & \square \\ \square & \square \\ \square & \square \\ \square & \square \\ \square & \square \\ \square & \square \\ \square & \square \\ \square & \square \\ \square & \square \end{array} \right.$$

By examination of this output it can be seen that these matrices form a group under matrix multiplication. They are 4x4 and real valued. As a group they are simple. This group also represents a continuous symmetry of the body centered cubic lattice and as such this is a lie group. Therefore this must be a real representation of the exceptional lie group F(4)/Spin(4). This will be the representation I will use in this composition.

The Lie Algebra F(4)/Spin(4)

Does F(4)/Spin(4) have an associated Lie Algebra?

A Lie algebra is a vector space L defined over the real numbers with a binary operation [,]that satisfies the following axioms. Where X and Y are in L.

$$[(X1 + X2), Y] = [X1, Y] + [X2, Y] \tag{0.22}$$

$$[(qX), Y] = q[X, Y] \quad q\ is\ a\ complex\ scalar \tag{0.23}$$

$$[X, Y] = -[Y, X] \tag{0.24}$$

$$[X, [Y, Z]] + [Z, [X, Y]] + [Y, [X, Z]] = 0 \tag{0.25}$$

Take as L the vector space of diagonal 4x4 matrices with the basis for the space being {η ,θ ,κ ,Θ}Those being 4x4 matrices with a 1 in one of the diagonal positions. Any diagonal matrix can be written in terms of those matrices. Then take the Lie bracket as being the commutator.

The fourth axiom is satisfied by the commutativity of all diagonal matrices as is axiom 3. The linearity of the commutator and matrices guarantees that the first and second axioms are also satisfied.

Because these axioms are satisfied I can say that this is a lie algebra. This empowers me to use F(4)/Spin(4) to investigate further.

The Gauge Field With the Symmetry of F(4)/Spin(4).

So what would a field with the gauge symmetry of F(4)/Spin(4) look like? What would a Lagrangian have to look like for it to be invariant under a F(4)/Spin(4) transformation?

Before I answer these questions I want to make an observation about the structure of F(4)/Spin(4). Any element of F(4)/Spin(4) can be written in terms of four of the matrices in the representation used in this paper. Specifically {η ,θ ,κ ,Θ}Taken together these can form a soft of pseudo four vector.

$$F_\lambda = a\eta + b\theta + c\kappa + d\lambda \tag{0.26}$$

This is true because all of the matrices in this representation are diagonal and they have one's or zero's on the diagonal. The elements of those four matrices in the representation of F(4)/Spin(4) have just a one on the diagonal and zero's elsewhere. Therefore they form a basis for the space of possible transformations in F(4)/Spin(4) as represented by these matrices. This construction would also have all the same algebra as the matrices in F(4)/Spin(4) do.

So what field would be symmetric under this transformation?

Consider the stress energy tensor $T_{\alpha\beta}$. Of all the fields in my theory it is the most physical and unambiguously defined.

$$T_{\alpha\beta} \to F_\lambda \, T_{\alpha\beta} \, F^{\lambda+} \tag{0.27}$$

Where the + super script indicates the pseudo inverse of the matrix. A pseudo inverse has to be used because these matrices are singular.

Working through this the field in my theory and the field of empty space-time in classical general relativity are diagonal. The fields F_λ are diagonal. Because of the high degree of symmetry that exist in diagonal matrices we can shuffle these fields around at will. So what I will do is re write the last equation like so.

$$T_{\alpha\beta} \to F_\lambda \, T_{\alpha\beta} \, F^{\lambda+} \to T_{\alpha\beta} \, F_\lambda \, F^{\lambda+} \tag{0.28}$$

By the usual rules of the pseudo inverse I could re write the last F's as just one F. Instead I will use the fact that a pseudo inverse in effect as well as einstein summation. Each of those λ 'sis a matrix and each of those matrices is a zero with a one at a diagonal. Therefore $F_\lambda \, F^{\lambda+} \to I$ or the identity matrix. Therefore I can write that

$$T_{\alpha\beta} \to F_\lambda \, T_{\alpha\beta} \, F^{\lambda+} \to T_{\alpha\beta} \, F_\lambda \, F^{\lambda+} \to T_{\alpha\beta} \, I \to T_{\alpha\beta} \tag{0.29}$$

Therefore any diagonal matrix representation of these fields would have F(4)/Spin(4) symmetry... However there is only one field that has to be diagonal in any basis for physical reasons. The stress energy tensor of general relativistic space-time. It's matrix rep has to have only diagonal entries just like that of an ideal fluid.

Therefore I say that this field of quantized space-time is THE field which is invariant under the F(4)/Spin(4) transformation. This may sound a bit odd to at first until one considers that the classical Einstein equation in a mass free region is a simple scaling relationship between the Einstein tensor and the stress energy tensor. $G^{\mu\nu} = 8\pi T^{\mu\nu}$.

The Lagrangian

The next step is to write the Lagrangian for this field. The question is this: what is the simplest lagrangian that can be made out of the given mathematical objects in this theory? First consider what the free field Lagrangian would look like.

In gravity what would happen is the non gravitational stress energies propagate through what they see as this locally flat space- time. But as they propagate along as the available geodetic paths' vary they are pulled along through to their possible futures. These possible futures would be given by the S tensor in my theory. $S^{\mu\nu}$ Therefore I propose that a term in this Lagrangian should be of the form $\Delta_\lambda \, S^{\alpha\beta} \, T_{\alpha\beta}$. ($\Delta_\lambda$ is a covariant quantum derivative see here for what that means) but this form has the problem of not being a Lorentz scalar. If I use the f matrices just as the γ 's are used in QED I can make it a Lorentz scalar.

$$L = f^\lambda \, \Delta_\lambda \, S^{\alpha\beta} \, T_{\alpha\beta} = f^\lambda \, \Delta_\lambda \, @^\alpha{}_\delta \, T^{\delta\beta} \, T_{\alpha\beta} \tag{0.30}$$

Now I will consider the interaction term. Thinking physically what happens in gravitation is one particle interacts with another particle by exchanging a graviton.   In this case I shall use the A tensor and the @ tensor.

$$L_{\text{int}} = -\sqrt{G} \; G \, A^{\alpha\delta} \, @_\delta{}^\gamma \, A_{\alpha\gamma} \tag{0.31}$$

Last but not least there is the fact that we have a field with an Abelian field with a lie algebra and it's symmetry. We know from experience with QED that a gauge invariant field needs to be introduced.

$$F_\alpha{}^{\delta\beta} = \left(\Delta_\alpha \, A^{\delta\beta} - \Delta_\delta \, A^{\alpha\beta} \right) \tag{0.32}$$

Now the total Lagrangian can be written.

$$L = f^\lambda \, \Delta_\lambda \, @^\alpha{}_\delta \, T^{\delta\beta} \, T_{\alpha\beta} - \sqrt{G} \; G \left(\frac{1}{4} \, F_\alpha{}^{\delta\beta} \, F^\alpha{}_{\delta\beta} - A^{\alpha\delta} \, @_\delta{}^\gamma \, A_{\alpha\gamma} \right) \tag{0.33}$$

Conserved Current

My next task is to elevate F(4)/Spin(4) to a local symmetry. This is not difficult. All one needs to do is consider a single element of the basis for the space F(4)/Spin(4). Like so

$T_{\alpha\beta} \to f \, T_{\alpha\beta} \, f^+ \to f \, f^+ \, T_{\alpha\beta} \to f \, T_{\alpha\beta}$

$T_{\alpha\beta} \to f \, T_{\alpha\beta}$ Is by the nature of this theory both discrete and continuous. I say this because this theory is formulated to be most valid at the Planck scale which theoretically is the smallest meaningful infinitesimal increment that can be. So to transform by such an increment meet's both definitions.

Like all continuous symmetries of the action this one will also have a conserved current. Figuring this out is elementary to QFT since the Lagrangian is known. The conserved Current is

$$J^{\alpha\beta} = f^{\lambda} \, @^{\alpha}{}_{\lambda} \, f_{\delta} T^{\delta\beta}$$

Looking at the dimensions this has the units of length of length and the operator $S^{\alpha\beta} = @^{\alpha}{}_{\lambda} T^{\lambda\beta}$ is embedded within. This current is of space-time and the conserved quantity is in fact space-time.

Application of Quantized Space - Time to Black Hole Physics.

The first consequence of the postulate of quantized space-time is that all physically meaningful lengths, areas, and volumes are quantized. The unit of quantization for reasons already established is the planck length-time, planck area, and planck volume. The planck mass is also a fundamental unit of mass as far a black hole physics is concerned, because a mass less than the planck mass cannot be collapsed onto itself. Here is a quantitative way of demonstrating this will be to put the Schwarzschild radius into a form which agree's with the postulate of quantized space-time.

The Schwarzschild Radius.

This is exactly how I realized that the Schwarzschild Radius can be found from this theory. Using an older and simpler definition of the operators. The graviton-area operator @, the Quantum energy distribution tensor **A** and the mass tensor **M**. How can I combine these to form useful quantities. So I multiplied these together to see if anything of interest came out. The most interesting result was that If I multiply @ by **M** I get a quantity with the dimensions of length.

$$H = M. \tag{0.34}$$

Expanding this quantity in any orthonormal basis gives a diagonal unit four tensor-operator times M.

$$M_{ij} = \begin{pmatrix} 1 & 0 & 0 & 0 \\ 0 & 1 & 0 & 0 \\ 0 & 0 & 1 & 0 \\ 0 & 0 & 0 & 1 \end{pmatrix} M \tag{0.35}$$

In this case due to the symmetry of a spherical object the @ operator can be simplified and all of the n_i set equal to one. Thus the operator becomes Newtons gravitational constant times the Minkowski metric. The product of these operators can be simply computed as they both are diagonal. The trace of this operator is.

$$\text{Tr}\left[\begin{pmatrix} -GM & 0 & 0 & 0 \\ 0 & GM & 0 & 0 \\ 0 & 0 & GM & 0 \\ 0 & 0 & 0 & GM \end{pmatrix} \right] \tag{0.36}$$

$$2\,G\,M \tag{0.37}$$

In the units chosen for this paper ($\hbar^2 = c^2 = 1$) Schwarzschild radius is well known to be [3].

$$2\,G\,M \tag{0.38}$$

This is exactly right. Now put the mass M in units of Planck mass and the expression becomes.

$$2\,\sqrt{G}\,\,m \tag{0.39}$$

$$m = \{1,\ 2,\ 3,\ \ldots\} \tag{0.40}$$

Which is basically the same as equation one.

The General Quantum Schwarzschild Radius Formula .

From the above simplified case of a point mass it is obvious that the quantum Schwarzschild radius formula gives the same results as the classical Schwarzschild radius up to one Planck Length. Next I demonstrate how this formula would work for a more general mass-energy distribution.

In[43]:= **FullSimplify$\left[\hat{at}.\hat{M}\right]$ // MatrixForm**

Out[43]//MatrixForm=

$$
\begin{pmatrix}
-\dfrac{G\sqrt{-n_0^3+n_1^3+n_2^3+n_3^3}}{2\,n_0\sqrt{G\,n_0\,n_1\,n_2\,n_3}} & \dfrac{G\,n_0\sqrt{-n_0^3+n_1^3+n_2^3+n_3^3}}{2\,n_1^2\sqrt{G\,n_0\,n_1\,n_2\,n_3}} & \dfrac{G\,n_0\sqrt{-n_0^3+n_1^3+n_2^3+n_3^3}}{2\,n_2^2\sqrt{G\,n_0\,n_1\,n_2\,n_3}} & \dfrac{G\,n_0\sqrt{-n_0^3+n_1^3+n_2^3+n_3^3}}{2\,n_3^2\sqrt{G\,n_0\,n_1\,n_2\,n_3}} \\[2em]
-\dfrac{G\,n_1\sqrt{-n_0^3+n_1^3+n_2^3+n_3^3}}{2\,n_0^2\sqrt{G\,n_0\,n_1\,n_2\,n_3}} & \dfrac{G\sqrt{-n_0^3+n_1^3+n_2^3+n_3^3}}{2\,n_1\sqrt{G\,n_0\,n_1\,n_2\,n_3}} & \dfrac{G\,n_1\sqrt{-n_0^3+n_1^3+n_2^3+n_3^3}}{2\,n_2^2\sqrt{G\,n_0\,n_1\,n_2\,n_3}} & \dfrac{G\,n_1\sqrt{-n_0^3+n_1^3+n_2^3+n_3^3}}{2\,n_3^2\sqrt{G\,n_0\,n_1\,n_2\,n_3}} \\[2em]
-\dfrac{G\,n_2\sqrt{-n_0^3+n_1^3+n_2^3+n_3^3}}{2\,n_0^2\sqrt{G\,n_0\,n_1\,n_2\,n_3}} & \dfrac{G\,n_2\sqrt{-n_0^3+n_1^3+n_2^3+n_3^3}}{2\,n_1^2\sqrt{G\,n_0\,n_1\,n_2\,n_3}} & \dfrac{G\sqrt{-n_0^3+n_1^3+n_2^3+n_3^3}}{2\,n_2\sqrt{G\,n_0\,n_1\,n_2\,n_3}} & \dfrac{G\,n_2\sqrt{-n_0^3+n_1^3+n_2^3+n_3^3}}{2\,n_3^2\sqrt{G\,n_0\,n_1\,n_2\,n_3}} \\[2em]
-\dfrac{G\,n_3\sqrt{-n_0^3+n_1^3+n_2^3+n_3^3}}{2\,n_0^2\sqrt{G\,n_0\,n_1\,n_2\,n_3}} & \dfrac{G\,n_3\sqrt{-n_0^3+n_1^3+n_2^3+n_3^3}}{2\,n_1^2\sqrt{G\,n_0\,n_1\,n_2\,n_3}} & \dfrac{G\,n_3\sqrt{-n_0^3+n_1^3+n_2^3+n_3^3}}{2\,n_2^2\sqrt{G\,n_0\,n_1\,n_2\,n_3}} & \dfrac{G\sqrt{-n_0^3+n_1^3+n_2^3+n_3^3}}{2\,n_3\sqrt{G\,n_0\,n_1\,n_2\,n_3}}
\end{pmatrix}
$$

Take the trace of this operator and we get the full quantum schwarzschild radius formula.

In[44]:= **FullSimplify$\left[\text{Tr}\left[\hat{at}.\hat{M}\right]\right]$**

Out[44]=
$$
\frac{G^2\sqrt{-n_0^3+n_1^3+n_2^3+n_3^3}\;\left(-n_1\,n_2\,n_3+n_0\,(n_2\,n_3+n_1\,(n_2+n_3))\right)}{2\,(G\,n_0\,n_1\,n_2\,n_3)^{3/2}}
$$

Black Hole Geometry.

Using the first approximation to the quantum Schwarzschild radius formula many useful quantities can be computed. In particular the area and volume of a quantum black hole.

In[55]:= $R_s = \dfrac{1}{2}\,\text{Tr}\left[\begin{pmatrix} -G & 0 & 0 & 0 \\ 0 & G & 0 & 0 \\ 0 & 0 & G & 0 \\ 0 & 0 & 0 & G \end{pmatrix}.\begin{pmatrix} -1 & -1 & -1 & -1 \\ 1 & 1 & 1 & 1 \\ 1 & 1 & 1 & 1 \\ 1 & 1 & 1 & 1 \end{pmatrix}\right]*\dfrac{n}{\sqrt{G}}$

Out[55]=

$2\sqrt{G}\,n$

What this formula tells us is for quantum gravitational purposes only scalar multiples of the planck energy-mass result in gravitational change. This total of energy is aware of all the energies of all forms in the space in question. The Planck mass-energy can be reached by supposedly empty space if it contains enough radiation. The quantization of the black hole's radius tells us that it's area and volume are also quantized. It would serve our purposes nicely to find an acceptable formula for the area, and volume of a black hole in quantum terms. If we apply the area operator to the mass-energy tensor of the black hole then multiply by a unit of planck length we get a quantity with the dimensions of area. It has the desirable characteristic that it does not contain pi.

In[2]:= $A_s = \text{Tr}\left[\begin{pmatrix} -G & 0 & 0 & 0 \\ 0 & G & 0 & 0 \\ 0 & 0 & G & 0 \\ 0 & 0 & 0 & G \end{pmatrix}.\begin{pmatrix} -1 & -1 & -1 & -1 \\ 1 & 1 & 1 & 1 \\ 1 & 1 & 1 & 1 \\ 1 & 1 & 1 & 1 \end{pmatrix}*\dfrac{n}{\sqrt{G}}*\sqrt{G}\right]$

Out[2]=
$4\,G\,n$

Now for the volume of a black hole. Apply the area operator twice to the mass tensor then take the trace of the resulting matrix.

In[1]:= $V_s = 2\,\text{Tr}\left[\begin{pmatrix} -G & 0 & 0 & 0 \\ 0 & G & 0 & 0 \\ 0 & 0 & G & 0 \\ 0 & 0 & 0 & G \end{pmatrix}.\begin{pmatrix} -G & 0 & 0 & 0 \\ 0 & G & 0 & 0 \\ 0 & 0 & G & 0 \\ 0 & 0 & 0 & G \end{pmatrix}.\begin{pmatrix} -1 & -1 & -1 & -1 \\ 1 & 1 & 1 & 1 \\ 1 & 1 & 1 & 1 \\ 1 & 1 & 1 & 1 \end{pmatrix}\right]*\dfrac{n}{\sqrt{G}}$

Out[1]=
$4\,G^{3/2}\,n$

You may have noticed that these formulae bear no resemblance to those used for a classical sphere. This is as it should be as they are not really spherical objects at the quantum level. They will be a sort of spherically symmetrical cloudy, uncertain object. The uncertainty principle plays a strange role in the fine geometry in quantum space-time. When we think geometry we think of sharp defined lines and exact computations. Quantum geometry will feature lines points and curves that are out of focus. Determining weather two lines intersect will simply be a probability of intersection.

Quantum Gravitational Collapse.

The classical process of gravitational collapse is what happens when a object collapses into a black hole. This ca be triggered by any number of processes. I will address the general subject of collapse then I will consider two possibilities. The first one will be a supernova resulting in black hole formation. The second will be a star loosing it's internal fusion then collapsing under its own gravity into a black hole. These are two different processes but they both have one critical thing in common. The density of the resulting object.

Density is defined as the mass per unit volume. If an object is to become a black hole then it must be compressed into a volume smaller than or equal to it's schwarzschild volume. The instant this object reaches this point it becomes a black hole. It does not matter how the object reaches that state as long as it is of planck mass or larger. Mathematically this formula would be trivial

In[3]:= $\rho_c = \dfrac{m}{V_s} = \mathrm{Tr}\left[\begin{pmatrix} -1 & -1 & -1 & -1 \\ 1 & 1 & 1 & 1 \\ 1 & 1 & 1 & 1 \\ 1 & 1 & 1 & 1 \end{pmatrix} n \middle/ \left(\sqrt{G}\right)\right] \middle/$

$\left(2\,\mathrm{Tr}\left[\begin{pmatrix} -G & 0 & 0 & 0 \\ 0 & G & 0 & 0 \\ 0 & 0 & G & 0 \\ 0 & 0 & 0 & G \end{pmatrix} \cdot \begin{pmatrix} -G & 0 & 0 & 0 \\ 0 & G & 0 & 0 \\ 0 & 0 & G & 0 \\ 0 & 0 & 0 & G \end{pmatrix} \cdot \begin{pmatrix} -1 & -1 & -1 & -1 \\ 1 & 1 & 1 & 1 \\ 1 & 1 & 1 & 1 \\ 1 & 1 & 1 & 1 \end{pmatrix}\right] * n \middle/ \left(\sqrt{G}\right)\right)$

Out[3]=

$\dfrac{1}{2\,G^2}$

From In[27]:=

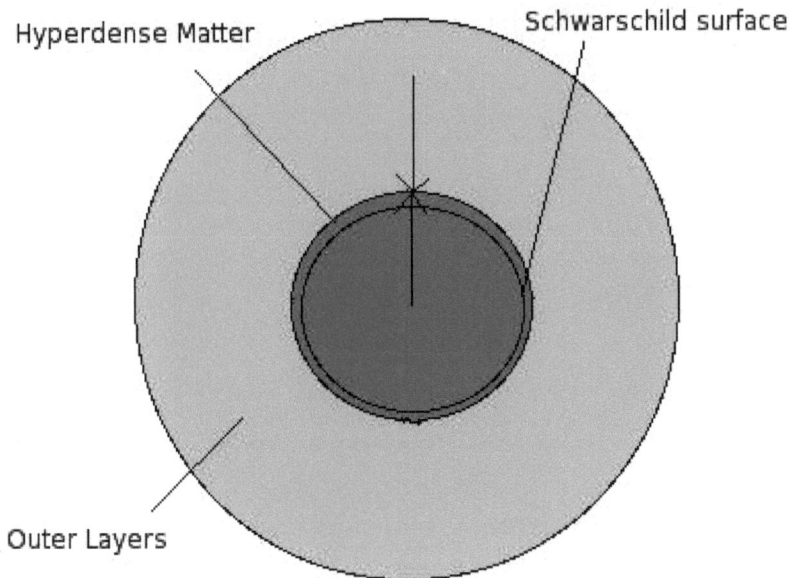

Hyperdense Matter

Schwarschild surface

Outer Layers

The illustration above is supposed to be the simplified inner structure of a star. The hyper dense matter at the center is where the fusion occurs. Outward force of the fusion is represented by the arrow pointing out. The inward force of the gravity of the star is the arrow pointing in. In a stable situation these forces balance and the star shines for billions of years. The force of gravity keeps the fusion contained and indeed is what starts it in the first place. Most stars are not massive enough to precipitate a black hole as they are not dense enough. Most stars simply shed their outer layers and leave behind a neutron star. Some objects which are not quite massive enough to form black holes for a class of object called a quark star. The defining difference between these objects is their density.

Stars that are the heaviest of the heavy result in supernovas. At the cores of these stars there is a region of hyper dense matter which is already at the brink of being compressed to the critical density. When the super nova occurs the explosion of the outer layers if it happens just right such that their is no relief of the pressure in any direction the hyperdense matter reaches or even exceeds the critical density and becomes an ultra-hyperdense object that we call a black hole. In this conception of a black hole their is no singularity at the center where physics breaks down. Instead their is just another state of matter hidden from view by the fact that no direct radiation may leave if it gets too close. This hyperdense sphereoid may just be one or two $\sqrt{G}$ beneath the event horizon or it may be much much farther below the event horizon. There is no way I can think of to know.

The other scenario would be for a star to simply collapse in on itself of it's own weight. Such a process is what most people envision as gravitational collapse. This is the case when say a small part of a black hole is near critical density even though the star is still shinning it is shinning less and less brightly. With less nuclear fusion to keep the star's core from collapsing due to it's weight and the weight of the upper layers the core collapses into a black hole. This state of affairs would not last long because the matter adjacent to the hole would fall in and the hole would grow. In increments of planck radius, mass, volume and area the hole would grow through a process of quasistatic steps. This continues until the hole's horizon reached the outer most layer of the star and the star seems to disappear. Having collapsed on itself.

The Classical Thermodynamics of Quantum Black Holes.

We have two widely accepted approaches to the thermodynamics of black holes. The first approach I have the most affinity for is the approach of Beckenstien which looks at classically scaled black holes through information theory. The second approach is the approach of Hawking which, through observation and thought experiments, establishes four laws of black hole mechanics. Then makes the leap by analogy to the laws of thermodynamics. First let us look at the laws of black hole mechanics. These laws are based on the results of numerous thought experiments backed by astronomical observations of known black holes.

Law zero:
The surface gravity (κ)electric potential (ϕ)and angular velocity of the event horizon(Ω)are all constant over the surface of the black hole.

The First Law:
The change in mass of a black hole is equal to the products of the surface gravity and change in area,angular velocity and change in angular momentum,and the electric potential and change in charge.
Mathematically

$$T \, dM = \frac{1}{8\pi} \kappa \, dA + \Omega \, dJ + \phi \, dq$$

(0.41)

The second law:
The area of a black hole cannot decrease in any interaction or transformation.

$$\Delta A \geq 0$$

(0.42)

The Third Law :
No finite number of processes can reduce the area of a black hole to zero.
Eyeballing these laws of black hole mechanics one who is versed in thermodynamics can see the similarities. This is part of the argument used to establish that the laws of thermodynamics apply to black holes by Hawking. Mathematically he used a set of transformations to show mathematically the process of black holes emitting particles by quantum mechanical means. Once he saw that black holes radiate like black bodies in thermodynamics only then did he accept the connection we now regard as obvious.

A Heuristic Example

Now let us look at black holes in context of information theory as Beckenstien did. Take a Schwarzschild black hole and a test assembly of particles. Let us assume that the ball of particles was a spinning electrostatically charged crystalline solid and perfectly pure. We have full and perfect knowledge of the states of every particle the ball is made of. What happens to the information we have about the states of that assembly of particles when the ball falls in the hole? Beckenstien says that the black hole's event horizon hides the information about what is inside the black hole. The key definition for entropy in terms of information is that the change in the amount of information is the negative of the change in enthropy. The information we are discussing here would be all the states of energy of the assembly of particles that went into the black hole. However, according to the no-hair theorem, the black hole does let us know the mass, net charge and angular momentum of it's constituents. When the spinning charged ball of particles fell into the hole it would grow in volume, and transform in to a Kerr-Newman black hole. The enthropy of the new system composed of a schwarzschild black hole will be the amount of information lost to the hole. This difference will be the change in enthropy of the system. The same formula should describe the engorgement of new matter-energy in to the black hole. The basic relationship between enthropy S, and information I is.

$$\Delta S = -\Delta I$$

(0.43)

Enthropy, Information and Heat of a Black Hole.

In a quantum theory of space-time right from the beginning we need to recognize that a statistical approach will most validly lead to correct results. Such a statistical approach will keep account of the quantum states knowable before collapse and those that are hidden by collapse. The challenge is to describe the information at all phases of the formation of the black hole. The theory of quantized space-time has provided that by identifying the critical quantity which defines a black hole, its density. Nothing special happens to space at the event horizon it is just a region of space filled with ultra-hyperdense matter-energy fields. The gravitational field of altered probability caused by them is no different than that of the matter which went in. This is a sticky place where classical and quantum effects are of almost equal importance. To keep the problem quantum mechanical let us restrict our query to black holes that are only a few planck masses in size. Let us assume that they are in thermal equilibrium with their surroundings so that any possible evaporation or explosion will not happen.

For the purposes of quantum mechanics the information is quantum numbers. The question is how many quantum numbers can a black hole hide? The maximum information encoded in a particle will be found by counting it's possible quantum numbers. Each and every particle and field will have a quantum number to describe it's mass-energy, charge, spin, strangeness, charm(ingness), topness, bottomness, upness and downness, lepton number, isospin, color, baryon number, parity, any number of bits of information that we do not know about. We will call the number of quantum numbers possible i. Classically or quantum mechanically as a particle collapses into a black hole we can only know M,Q, and L. So the change in information will be

$$\Delta\ I= i - 3 \tag{0.44}$$

for a particle that has collapsed on iteslf into a black hole. This would happen if a single quantum were to be accelerated untill its total energy was at least equal to the planck energy (about 3.8x10^26 eV).

How about the case of a black hole that has existed for some time. The hole is large and cool and relatively stable over reasonable periods of time regardless of the surroundings. We already know what the information loss per quantum particle is Δ I=i-3. What we need to know now is the maximum particles,quanta of field can be in a black hole at maximum. Each particle or field must have a certain spatial extent. They need a certain volume to exist in. At minimum each particle or field quantum could be localized to a planck scale hyper-cube. The number of particles that will be contained within the hole (N) will be found by dividing the volume of the hole by the planck volume.

$$\mathrm{In[1]:=} \quad N = \frac{V_s}{V_p} = \frac{4\,G^{3/2}\,n}{G^{3/2}}$$

Out[1]=

4 n

This means that the number of particles or fields the hole can hold in will be four times the number (n) of planck mass-energies that went into making the hole. That means a minimal one planck energy hole could only hold in four quantum particles. The enthropy of a black hole would be.

$$\Delta\ S= N\Delta\ I= N\,(i - 3) = 4\,n\,(i - 3) \tag{0.45}$$

The prime test of a theory of quantum gravity seems to be the relationship between thermodynamic, geometric, and gravitational quantities. This last equation shows a relationship between the change in entropy of a black hole, the volume of the black hole, and it's mass. The instantaneious entropy of a black hole will be computed from the usual means of Boltzmans theorem S=klogω . Where ω will be the distribution of particles over states known as the thermodynamic probability. In a black hole the particles could be Fermions or Boseons and their is no way we can distinguish between the two. So the system will be partly described by Fermi-Dirac statistics and partly by Bose-Einstien Statistics. Next let us consider the number of particles in a black hole of one unit of planck mass. The smallest theoretically possible black hole. N=4n where n=$E\sqrt{G}$ and E is the total mass energy of the hole as given previously in the paper. If E=1 then N=4 particles confined to the hole by gravity alone! So the rest of the mass in the hole is liable to be lost as radiation unless the environment is hot enough to restore some sort of thermal equilibrium. The matter in the hole is very dense but it is in a very fluid state with old energy leaving and new energy entering at all times. The matter which is stably in the hole is responsible for absorbing then radiating this energy so what we have is a very few particles which are constantly switching through a much larger number of states in a system where their identities as fermions and boseons cannot be determined. This describes a system ruled by Maxwell-Boltzmann statistics.

$$\omega = \prod_{j=1}^{m} \frac{Q_j^{N_j}}{N_j!}\ ;\ j = 1,\,2,\,3,\,m^{\mathrm{th}}\ \text{quantum state.} \tag{0.46}$$

Before you the scientific reader hrumph and reject this remember the area,volume etc of the hole is encoded in the number of particles it can hold N.So the enthropy of a black hole depends on the area of the hole in this theory in just the way we would expect it should.

$$S = k\log\Big[\prod_{j=1}^{m} \frac{Q_j^{N_j}}{N_j!}\Big] \tag{0.47}$$

Fin the temperature of the hole from a classic equation

$$du = Tds. \tag{0.48}$$

Treat T as constant then take the indefinite integral of both sides.

$$U = TS == > T = \frac{U}{S}, \quad T = U \bigg/ k \text{Log}\bigg[\prod_{j=1}^{m} \frac{Q_j^{N_j}}{N_j!}\bigg] \qquad (0.49)$$

In the last Line U would be the internal energy of the system but not necessarily incriments of planck mass. In other words the temperature of the hole can undergo small fluctuations and not cause gravitational consequences if

$$\Delta U < \frac{n}{\sqrt{G}} \, . \qquad (0.50)$$

Gravitational Collapse as a Thermodynamic Process.

In this section I will describe the collapse of a star into a black hole as a thermodynamic process. Thermodynamics coupled with quantum geometry will provide a mathematical description of this process. The crucial details will be illustrated by considering a schwarzschild black hole .

For such a black hole we have the relation to describe the temperature enthropy and internal energy of a state. This formula will allow a mathematical description of gravitational collapse. The way the formula is written assumes that the change in internal energy is relatively small compared to the planck mass so the number of particles that can be confined by the hole remains steady. As a black hole expands or collapses that number will change. To accomplish all of this replace U with E = n/squrt[G]. Then replace N with 4 n. The resulting formula will describe the collapse of a black hole in thermodynamic terms.

In[13]:=

$$T = \frac{U}{S} = \frac{\frac{n}{\sqrt{G}}}{k * \text{Log}\left[\prod_{j=1}^{m} \frac{Q_j^{(4n)_j}}{(4n)_j!}\right]}$$

Note how the enthropy in the denominator depends on the energy in the numerator via the index n. To get a preliminary feel for how this would look examine the graph below. This would roughly be the shape of a plot of the temperature as a function of the energy of the hole. Note how the temperature of the hole drops as (x which is standing in for) the mass increases. This is in line with the predictions that a black hole would radiate like a black body, and that as the mass increases the temperature decreases..

In[11]:=

$$\text{Plot}\left[\frac{4\,x}{\text{Log}\left[\frac{10^{4\,x}}{4\,x}\right]}, \{x, 0, 10\}\right]$$

From In[11]:=

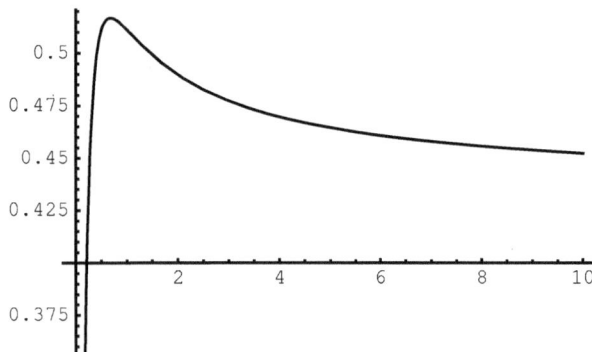

The formula I have derived so far is only valid for a Schwarzschild black hole but generalizing it to cover Kerr – Newmann black holes is no big deal. From the first law of black hole mechanics.

$$dM = \frac{1}{8\,\pi}\,\kappa\, dA + \Omega\, dJ + \phi\, dq \qquad (0.51)$$

If dM is the energy of the hole and $\frac{1}{8\,\pi}\,\kappa\, dA$ s the enthropy this can be re written .

$$\frac{U - (\Omega J + \phi q)}{S} = T \rightarrow T = \frac{\frac{n}{\sqrt{G}} - (\Omega J + \phi q)}{k \text{Log}\left[\prod_{j=1}^{m} \frac{Q_j^{N_j}}{N_j!}\right]} \tag{0.52}$$

With this last line we have a quantum mechanically correct candidate description of black hole collapse which does not run counter to any principle of relativity. This is so because the description does not make reference of space-time coordinates of any kind. Therefore the formulation invariant under any transformation of space-time coordinates.

Comparison to Known Experimental Data.

Theory is nothing without experiment. Theory can lead the way to an extent but most of the time theories are built from the bottom up. A theorist is presented with data and tries to find a physical-mathematical principle or equations that can predict the outcome of the experiment. The theory in this book is quite different just by nature. This is a theory which has been built from the top down. I started with an assumption or postulate about the fine scale structure of space time. Then from there I used mathematics and *Mathematica* to figure out what the consequences of that assumption are. From there I have been able to make some more abstract predictions.

What I will do in this chapter is give some calculations that relate to experiments which have been done in the field of quatum gravity. One is the fall of neutrons in Earth's gravity, the other has to do with the energy needed to keep a satellite at altitude (excluding it's orbital motion around the planet.) The results are in one case right on the money, in the other very promising.

One gravitational field we can easily and cheaply experiment with is that of the planet earth. Every time we put a satellite in orbit we do an experiment with this gravitational field. One particular experiment has been done using the gravitational field of the earth and a beam of neutrons. I will describe how my theory can be used to explain these results in detail.

The Fall of a Neutron.

In [8], Nesvizhevsky et.al. Measured the eigenvalues of a beam of ultra cold neutrons falling in earth's gravitational field. The following is inspired by this experiment. In the experiment Nesvizhevsky and his team set up a ultra cold neutron source a neutron absorber and a mirror. As figure one shows.

For this part of the composition I will switch back to SI units for these calculations.

From In[25]:=

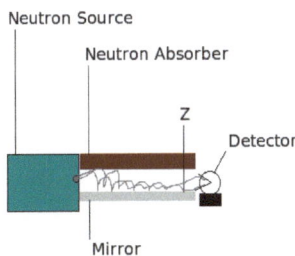

The only potential acting in this experiment is gravitational and in the vertical direction. In this case the forces are all conservative. With conservative forces the Hamiltonian is equal to the total energy.

$$H = c \, |p_z| - \frac{1}{2 \, n_z \, G} \, U \, P \, z^2$$

Remember in this equation p and P are not the same. P is the scalar relative momentum of the source of the gravitational field as seen from the test particle. U is the velocity of the earth as seen from the particle.

Substitute the momentum operator for p but not P. This gives the Hamiltonian operator.

$$\hat{H} = i \, c \, \hbar \, \partial_z - \frac{1}{n_z \, 2 \, G} \, U \, P \, z^2$$

The problem is to find an orthonormal basis that preferably gives H in a diagonal form. This will make life simpler in the future. Solving the problem in this way should give a full configuration interaction (CI), an exact solution to the problem. I propose the following set of basis vectors.

In[1]:= $\varphi_1 = \left\{ \frac{\sqrt{2}}{\sqrt{h}} \, \text{Sin}\left[\frac{z \, \pi}{h} \right] \right\}$

Out[1]= $\left\{ \frac{\sqrt{2} \, \text{Sin}\left[\frac{\pi \, z}{h} \right]}{\sqrt{h}} \right\}$

In[2]:= $\overline{\varphi_1} = \left\{ \dfrac{\sqrt{2}}{\sqrt{h}} \, \text{Sin}\left[\dfrac{z\,\pi}{h}\right] \right\}$

Out[2]= $\left\{ \dfrac{\sqrt{2}\ \text{Sin}\left[\frac{\pi z}{h}\right]}{\sqrt{h}} \right\}$

In[3]:= $\varphi_2 = \left\{ \dfrac{\sqrt{2}}{\sqrt{h}} \, \text{Cos}\left[\dfrac{z\,\pi}{h}\right] \right\}$

Out[3]= $\left\{ \dfrac{\sqrt{2}\ \text{Cos}\left[\frac{\pi z}{h}\right]}{\sqrt{h}} \right\}$

In[4]:= $\overline{\varphi_2} = \left\{ \dfrac{\sqrt{2}}{\sqrt{h}} \, \text{Cos}\left[\dfrac{z\,\pi}{h}\right] \right\}$

Out[4]= $\left\{ \dfrac{\sqrt{2}\ \text{Cos}\left[\frac{\pi z}{h}\right]}{\sqrt{h}} \right\}$

Where h and z are pure numbers nondimensionalized by division by the Planck Length. A quick test of orthonormality.

In[5]:= $\displaystyle\int_0^h \overline{\varphi_1} \cdot \varphi_2 \; dz$

Out[5]= 0

In[6]:= $\displaystyle\int_0^h \overline{\varphi_2} \cdot \varphi_2 \; dz$

Out[6]= 1

Out[123]=
1

Using this orthonormal basis the matrix elements can be computed.

In[7]:= $\begin{pmatrix} \int_0^h \overline{\varphi_1} \cdot \left(\mathbb{i}\,c\,\hbar\,\partial_z\,\varphi_1 - \dfrac{1}{2\,n\,G}\,U\,P\,z^2\,\varphi_1 \right) dz & \int_0^h \overline{\varphi_1} \cdot \left(\mathbb{i}\,c\,\hbar\,\partial_z\,\varphi_2 - \dfrac{1}{2\,n\,G}\,U\,P\,z^2\,\varphi_2 \right) dz \\ \int_0^h \overline{\varphi_2} \cdot \left(\mathbb{i}\,c\,\hbar\,\partial_z\,\varphi_1 - \dfrac{1}{2\,n\,G}\,U\,P\,z^2\,\varphi_1 \right) dz & \int_0^h \overline{\varphi_2} \cdot \left(\mathbb{i}\,c\,\hbar\,\partial_z\,\varphi_2 - \dfrac{1}{2\,n\,G}\,U\,P\,z^2\,\varphi_2 \right) dz \end{pmatrix}$ // **MatrixForm**

Out[7]//MatrixForm=

$\begin{pmatrix} -\dfrac{h^2\,P\left(2-\frac{3}{\pi^2}\right)U}{12\,G\,n} & \dfrac{h^2\,P\,U}{4\,G\,n\,\pi} - \dfrac{i\,c\,\pi\,\hbar}{h} \\[2ex] \dfrac{h^2\,P\,U}{4\,G\,n\,\pi} + \dfrac{i\,c\,\pi\,\hbar}{h} & -\dfrac{h^2\,P\left(2+\frac{3}{\pi^2}\right)U}{12\,G\,n} \end{pmatrix}$

The eigenvalues and eigenvectors of this system will be the compleat exact solution of this system.

In[8]:= {spectrum, states} = Eigensystem$\left[\begin{pmatrix} -\dfrac{h^2\,P\left(2-\frac{3}{\pi^2}\right)U}{12\,G\,n} & \dfrac{h^2\,P\,U}{4\,G\,n\,\pi} - \dfrac{i\,c\,\pi\,\hbar}{h} \\ \dfrac{h^2\,P\,U}{4\,G\,n\,\pi} + \dfrac{i\,c\,\pi\,\hbar}{h} & -\dfrac{h^2\,P\left(2+\frac{3}{\pi^2}\right)U}{12\,G\,n} \end{pmatrix}\right]$

Out[8]= $\left\{\left\{\dfrac{-2\,G\,h^4\,n\,P\,\pi^4\,U - 3\sqrt{G^2\,h^8\,n^2\,P^2\,\pi^4\,U^2 + G^2\,h^8\,n^2\,P^2\,\pi^6\,U^2 + 16\,c^2\,G^4\,h^2\,n^4\,\pi^{10}\,\hbar^2}}{12\,G^2\,h^2\,n^2\,\pi^4}, \right.\right.$

$\left.\dfrac{-2\,G\,h^4\,n\,P\,\pi^4\,U + 3\sqrt{G^2\,h^8\,n^2\,P^2\,\pi^4\,U^2 + G^2\,h^8\,n^2\,P^2\,\pi^6\,U^2 + 16\,c^2\,G^4\,h^2\,n^4\,\pi^{10}\,\hbar^2}}{12\,G^2\,h^2\,n^2\,\pi^4}\right\},$

$\left\{\left\{-\dfrac{i\left(G\,h^4\,n\,P\,\pi^2\,U - \sqrt{G^2\,h^8\,n^2\,P^2\,\pi^4\,U^2 + G^2\,h^8\,n^2\,P^2\,\pi^6\,U^2 + 16\,c^2\,G^4\,h^2\,n^4\,\pi^{10}\,\hbar^2}\right)}{G\,h\,n\,\pi^3\left(-i\,h^3\,P\,U + 4\,c\,G\,n\,\pi^2\,\hbar\right)}, 1\right\},\right.$

$\left.\left\{-\dfrac{i\left(G\,h^4\,n\,P\,\pi^2\,U + \sqrt{G^2\,h^8\,n^2\,P^2\,\pi^4\,U^2 + G^2\,h^8\,n^2\,P^2\,\pi^6\,U^2 + 16\,c^2\,G^4\,h^2\,n^4\,\pi^{10}\,\hbar^2}\right)}{G\,h\,n\,\pi^3\left(-i\,h^3\,P\,U + 4\,c\,G\,n\,\pi^2\,\hbar\right)}, 1\right\}\right\}\right\}$

Notice that the spectrum of this system has two distinctive eigenvalues indexed by "n". "n" would be the de Broglie wavelength of the neutron in a particular state. Taking the postulate of quantized space-time into account, "n" will be an integer multiple of the Planck length. "n" in this formula would be the number of planck length's long the de Broglie wavelength of the particle is at the given height. Obviously the minimum wavelength would be determined by the "rest" or ground state wavelength of the particle. In this experiment which eigenvalues are observed will depend on the altitude the experiment is performed at.

Plugging in integers one through infinity will give the various energies that the neutron would have in the gravitational well of the Earth. The lowest energy level will represent the energy required to lift a neutron out of Earth's gravitational well. That said the Hamiltonian operator solved in this example is not the most general. The potential energy formula used is only good for relatively small distances, for a system where gravity is the dominant interaction.

Numerical Analysis

To put numbers to these eigenvalues and eigenvectors raw numbers and physical constants have to be fed to the computer algebra system. I will use a conversion factor between joules and electron volts.

In[11]:= evj = 6 241 506 479 963 234 304 $\dfrac{\text{ev}}{\text{Joule}}$

In[51]:= ekg = 5.60959×10^{70}

I need the number of Planck length's long the neutron's ground state deBroglie wavelength is.

In[52]:=
c = 299 792 458

In[53]:=
n = $\left(\dfrac{\left(4.135667 * 10^{-15}\right)}{939 * 10^9}\right)\dfrac{1}{1.616 * 10^{-35}}$

In[54]:=
G = $6.67 \times 10^{-11} * \text{ekg}^{-1}$

In[55]:=
U = $\dfrac{5.001}{c}$

In[68]:=
m = $939.56536081 * 10^6$

In[70]:=
P = m U

In[58]:=
$\hbar = 6.5821191556 * 10^{-16}$

In[59]:=

$$h = 1.3 \times 10^{-11}$$

This will give the energy needed to boost a neutron from its ground state out of the earth's gravitational potential well.

In[71]:=

$$\left(\frac{-2\,G\,h^4\,n\,P\,\pi^4\,U - 3\,\sqrt{G^2\,h^8\,n^2\,P^2\,\pi^4\,U^2 + G^2\,h^8\,n^2\,P^2\,\pi^6\,U^2 + 16\,c^2\,G^4\,h^2\,n^4\,\pi^{10}\,\hbar^2}}{12\,G^2\,h^2\,n^2\,\pi^4} \text{ ev} \right)$$

Out[71]=

$$-3.41116 \times 10^{43} \text{ ev}$$

Using a classical formula I can estimate what altitude this should correspond to. This will give the height the neutron would have to rise in meters.

In[75]:=

$$\frac{3.41116 * 10^{43} \text{ ev} \frac{1}{\text{evj}}}{m\,(9.80)\,\text{Joule}}$$

Out[75]=

$$5.93553 \times 10^{14}$$

For comparison a light year is $9.4605284 * 10^{15}$ meters.

A neutron would have to be 593 Billion kilometers away from the Earth in order to be completely free of earth's influence. However at that altitude the energy levels will be so finely spaced that the motion of the neutron will be virtually free motion.

I cannot say anything definite about the experiment unless I know the height of the table or workbench that the experiment was performed on. I could guess that the height is about one meter.

In[227]:=

$$n1 = 1.6162412 * 10^{35}$$

In[228]:=

$$n2 = n1 + \left(1 * 10^{18}\right)$$

In[229]:=

$$\left(\frac{1}{12\,G^2\,h^2\,(n1)^2\,\pi^4} \right.$$
$$\left(-2\,G\,h^4\,(n1)\,P\,\pi^4\,U - 3\,\sqrt{G^2\,h^8\,(n1)^2\,P^2\,\pi^4\,U^2 + G^2\,h^8\,(n1)^2\,P^2\,\pi^6\,U^2 + 16\,c^2\,G^4\,h^2\,(n1)^4\,\pi^{10}\,\hbar^2} \right)$$
$$\text{ev} \right) - \left(\frac{1}{12\,G^2\,h^2\,(n2)^2\,\pi^4} \right.$$
$$\left(-2\,G\,h^4\,(n2)\,P\,\pi^4\,U - 3\,\sqrt{G^2\,h^8\,(n2)^2\,P^2\,\pi^4\,U^2 + G^2\,h^8\,(n2)^2\,P^2\,\pi^6\,U^2 + 16\,c^2\,G^4\,h^2\,(n2)^4\,\pi^{10}\,\hbar^2} \right) \text{ev} \right)$$

Out[229]=

$$0. \text{ ev}$$

Due to the limitations of my computer this is a close as it will come to stating the order of magnitude of the difference between two energy levels at this height. in spite of these limitations I can narrow down how high the table must be. The the n1 I choose corresponds to one meter. The amount added to n1 would be the margin of error for this estimate. So I would put the height of the table that the experiment was done on at about 1 meter give or take 10^{-17} meters. Of course the actual strength of gravitations pull on an object would depend on the local energy momentum density. For example the earth's gravitational field is stronger where the ground is denser and weaker a high altitudes. Laboratories wishing to repeat the experiments referred to need to consider such Geological factors when comparing their results.

The Nesvizhevsky experiment.

To analyze the experiment of Nesvizhevsky et al I find that it is helpful to switch from the Heisenberg form of the problem to the Schrodinger form of the problem. What I will have *Mathematica* do is solve the wave equation then plot the results. The wave equation will be derived from the Hamiltonian used in the preceding section.
The wave equation is then

$$\partial_{x,x}\,\psi[x] - k * \partial_x \left(x^2 * \psi[x]\right) + k^2 * x^4 * \psi[x] - m * \psi[x] == 0$$
$$-m\,\psi[x] + k^2\,x^4\,\psi[x] - k\left(2\,x\,\psi[x] + x^2\,\psi'[x]\right) + \psi''[x] == 0$$

Let's see what mathematica will do with it.

$\text{DSolve}\left[\left\{\partial_{x,x}\psi[x] - k*\partial_x\left(x^2*\psi[x]\right) + k^2*x^4*\psi[x] - m*\psi[x] == 0, \psi[0] == 0, \psi[h] == \text{Cos}[x]\right\}, \psi[x], x\right]$

$\text{DSolve}\left[\left\{-m\psi[x] + k^2 x^4 \psi[x] - k\left(2x\psi[x] + x^2\psi'[x]\right) + \psi''[x] == 0, \psi[0] == 0, \psi[h] == \text{Cos}[x]\right\}, \psi[x], x\right]$

Next I will try breaking it into it's linear components then recombining them.

$\text{DSolve}\left[\left\{\partial_{x,x}\psi[x] + x^4*\psi[x] == 0, \psi[0] == 0, \psi[h] == h^2\right\}, \psi[x], x\right]$

$$\left\{\left\{\psi[x] \to \frac{h^{3/2}\sqrt{x}\ \text{BesselJ}\left[\frac{1}{6}, \frac{x^3}{3}\right]}{\text{BesselJ}\left[\frac{1}{6}, \frac{h^3}{3}\right]}\right\}\right\}$$

$\text{DSolve}\left[\left\{\partial_{x,x}\psi[x] - \partial_x x^2*\psi[x] == 0, \psi[0] == 0, \psi[h] == h^2\right\}, \psi[x], x\right]$

$$\left\{\left\{\psi[x] \to \frac{\sqrt{3}\ h^2\ \text{AiryAi}\left[2^{1/3}x\right] - h^2\ \text{AiryBi}\left[2^{1/3}x\right]}{\sqrt{3}\ \text{AiryAi}\left[2^{1/3}h\right] - \text{AiryBi}\left[2^{1/3}h\right]}\right\}\right\}$$

So the total analytical general solution.

This solution will work for quantum objects (particles and fields) which are near each other. At longer distances the potential looks more like that described by general relativity.

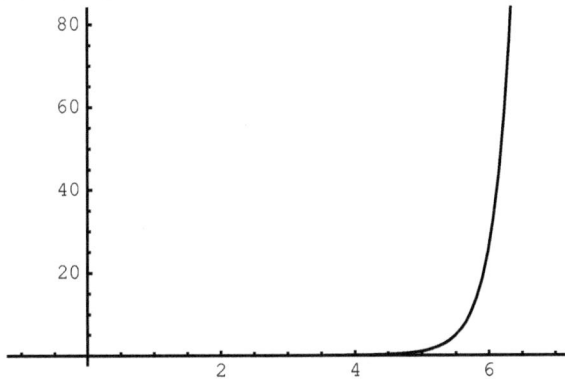

- Graphics -

$\text{DSolve}\left[\left\{\partial_{x,x}\psi[x] + x^4*\psi[x] == 0, \psi[0] == 6^2, \psi[6] == 6^2\right\}, \psi[x], x\right]$

$$\left\{\left\{\psi[x] \to \left(6^{5/6}\sqrt{x}\left(6^{2/3}\text{BesselJ}\left[\frac{1}{6}, \frac{x^3}{3}\right] + 6\ \text{BesselJ}\left[-\frac{1}{6}, \frac{x^3}{3}\right]\text{BesselJ}\left[\frac{1}{6}, 72\right]\text{Gamma}\left[\frac{5}{6}\right] - \right.\right.\right.\right.$$

$$\left.\left.\left.\left. 6\ \text{BesselJ}\left[-\frac{1}{6}, 72\right]\text{BesselJ}\left[\frac{1}{6}, \frac{x^3}{3}\right]\text{Gamma}\left[\frac{5}{6}\right]\right)\right)\middle/ \text{BesselJ}\left[\frac{1}{6}, 72\right]\right\}\right\}$$

$\{\{\psi[x] \to\}\}$

$\text{DSolve}\left[\left\{\partial_{x,x}\psi[x] - \partial_x x^2*\psi[x] == 0, \psi[0] == 0, \psi[6] == 6^2\right\}, \psi[x], x\right]$

$$\left\{\left\{\psi[x] \to \frac{36\left(\sqrt{3}\ \text{AiryAi}\left[2^{1/3}x\right] - \text{AiryBi}\left[2^{1/3}x\right]\right)}{\sqrt{3}\ \text{AiryAi}\left[6\ 2^{1/3}\right] - \text{AiryBi}\left[6\ 2^{1/3}\right]}\right\}\right\}$$

$\{\{\psi[x] \to\}\}$

$$\text{Plot}\left[\frac{36\left(\sqrt{3}\ \text{AiryAi}\left[2^{1/3}\,x\right] - \text{AiryBi}\left[2^{1/3}\,x\right]\right)}{\sqrt{3}\ \text{AiryAi}\left[6\,2^{1/3}\right] - \text{AiryBi}\left[6\,2^{1/3}\right]} + \right.$$

$$\left(6^{5/6}\sqrt{x}\,\left(6^{2/3}\,\text{BesselJ}\left[\frac{1}{6},\ \frac{x^3}{3}\right] + 6\,\text{BesselJ}\left[-\frac{1}{6},\ \frac{x^3}{3}\right]\,\text{BesselJ}\left[\frac{1}{6},\ 72\right]\,\text{Gamma}\left[\frac{5}{6}\right] - \right.\right.$$

$$\left.\left.6\,\text{BesselJ}\left[-\frac{1}{6},\ 72\right]\,\text{BesselJ}\left[\frac{1}{6},\ \frac{x^3}{3}\right]\,\text{Gamma}\left[\frac{5}{6}\right]\right)\right)\Big/\ \text{BesselJ}\left[\frac{1}{6},\ 72\right],\ \{x,\ 0,\ 7.5\}\right]$$

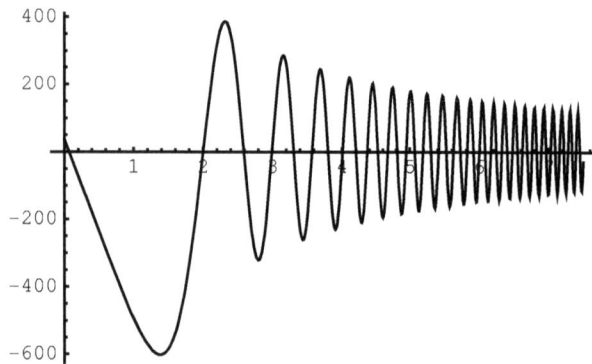

- Graphics -

In the following plot the probability is well behaved until the altitude of the neutron absorber in the experiment is reached. After that point the neutrons are basically free particles.

$$\text{Plot}\left[\left(\frac{36\left(\sqrt{3}\ \text{AiryAi}\left[2^{1/3}\,x\right] - \text{AiryBi}\left[2^{1/3}\,x\right]\right)}{\sqrt{3}\ \text{AiryAi}\left[6\,2^{1/3}\right] - \text{AiryBi}\left[6\,2^{1/3}\right]} + \frac{6\sqrt{6}\ \sqrt{x}\ \text{BesselJ}\left[\frac{1}{6},\ \frac{x^3}{3}\right]}{\text{BesselJ}\left[\frac{1}{6},\ 72\right]}\right)^2,\ \{x,\ 0,\ 7.5\}\right]$$

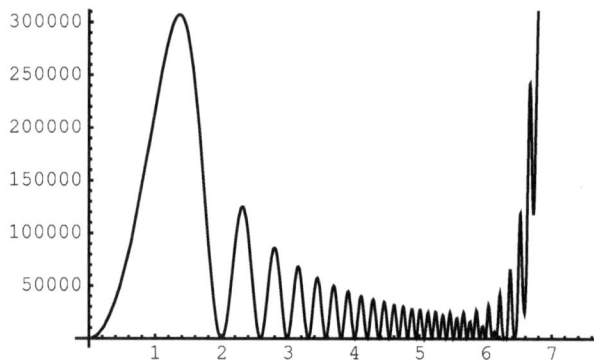

- Graphics -

The following is a plot of the potential surface.

```
Plot3D[y * x², {y, 0, 20}, {x, 0, 20}, ViewPoint -> {2.832, -1.753, 0.597}]
```

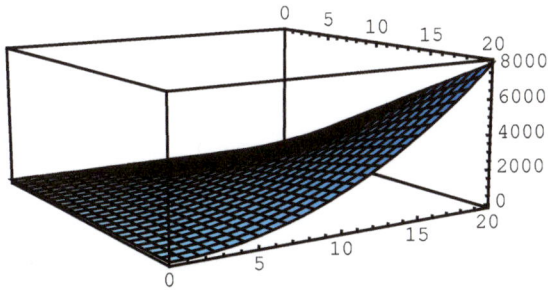

- SurfaceGraphics -

Now consider the problem with the upper boundary at infinity.

$$\text{DSolve}\left[\left\{\partial_{x,x}\psi[x] + x^4 * \psi[x] == 0, \psi[0] == 1, \psi[\infty] == 0, \psi'[0] == 0\right\}, \psi[x], x\right]$$

$$\left\{\left\{\psi[x] \to \frac{\sqrt{x} \ \text{BesselJ}\left[-\frac{1}{6}, \frac{x^3}{3}\right] \text{Gamma}\left[\frac{5}{6}\right]}{6^{1/6}}\right\}\right\}$$

$$\text{DSolve}\left[\left\{\partial_{x,x}\psi[x] - \partial_x x^2 * \psi[x] == 0, \psi[0] == 1, \psi[\infty] == 0\right\}, \psi[x], x\right]$$

{}

$$\text{Plot}\left[\left(\frac{\sqrt{x} \ \text{BesselJ}\left[-\frac{1}{6}, \frac{x^3}{3}\right] \text{Gamma}\left[\frac{5}{6}\right]}{6^{1/6}}\right)^2, \{x, 1, 5\}\right]$$

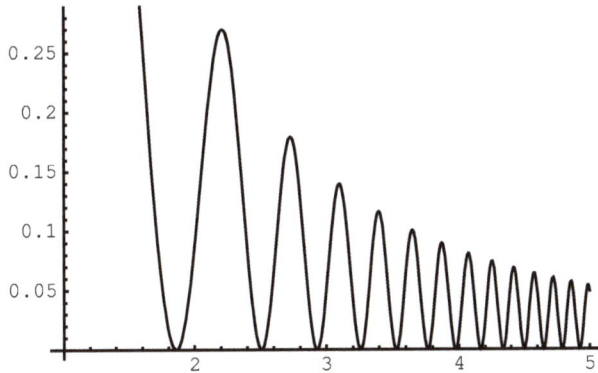

- Graphics -

$$\text{Plot}\left[\left(\frac{1}{6\ 6^{1/6}\ \sqrt{x}}\ \left(6^{1/3}\ \text{BesselJ}\left[\frac{1}{6},\ \frac{x^3}{3}\right]\ \text{Gamma}\left[\frac{1}{6}\right]+6\ x\ \text{BesselJ}\left[-\frac{1}{6},\ \frac{x^3}{3}\right]\ \text{Gamma}\left[\frac{5}{6}\right]\right)\right)^2,\ \{x,\ 0,\ 4.727\}\right]$$

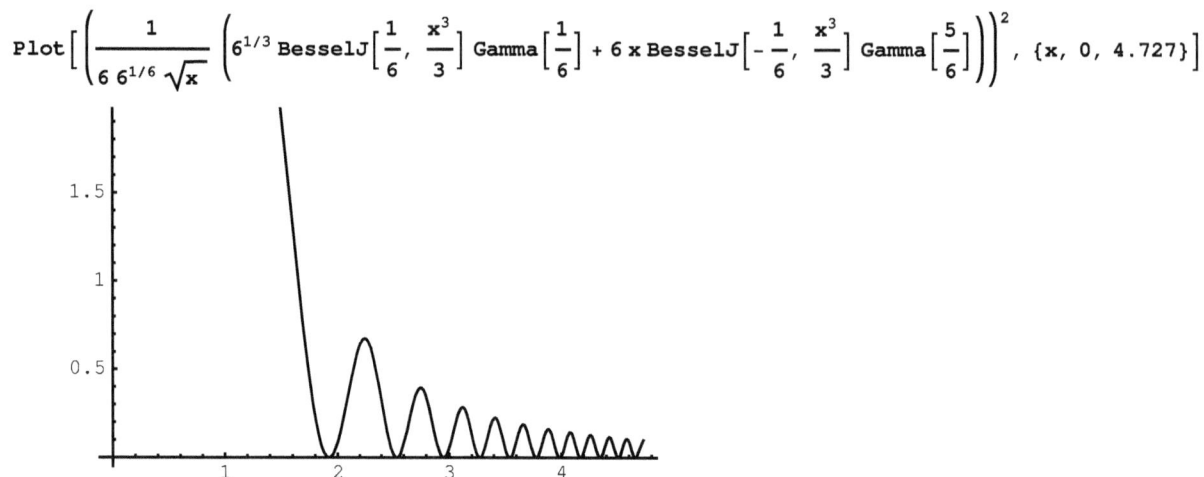

Notice how fast the energy drops off as the height increases along the x axis. This is as one would expect with the highest probability that the neutron will be found lower to the ground.

The energy required to boost a satellite to orbit.

Boosting a satellite into orbit is a problem more easily handled using a classical approach. However any quantum theory of gravity must be able to handle classical situations and give results similar to the classical theories. First to derive the Hamiltonian matrix for this type of system. The symmetry of this problem makes it effectively one dimensional.

In[13]:=
$$\psi_1 = \left\{\frac{1}{\sqrt{r}}\ \text{Exp}\left[i\ \frac{\xi\ \pi}{r}\right]\right\}$$

In[14]:=
$$\overline{\psi_1} = \left\{\frac{1}{\sqrt{r}}\ \text{Exp}\left[-i\ \frac{\xi\ \pi}{r}\right]\right\}$$

In[240]:=
$$\psi_2 = \left\{\frac{1}{\sqrt{r}}\ \text{Exp}\left[-i\ \frac{\xi\ \pi}{r}\right]\right\}$$

In[241]:=
$$\overline{\psi_2} = \left\{\frac{1}{\sqrt{r}}\ \text{Exp}\left[i\ \frac{\xi\ \pi}{r}\right]\right\}$$

Check the orthonormality.

In[15]:=
$$\int_0^r \overline{\psi_1}\ .\ \psi_1\ d\xi$$

Out[242]=
1

In[243]:=
$$\int_0^r \overline{\psi_1}\ .\ \psi_2\ d\xi$$

Out[243]=
0

The matrix elements of the kinetic energy are.

In[244]:=
$$\begin{pmatrix} \int_0^r \overline{\psi_1}\ .\ \left(\left(i\ c\ \hbar\ \partial_\xi\ \psi_1\right)\right)\ d\xi & \int_0^r \overline{\psi_1}\ .\ \left(\left(i\ c\ \hbar\ \partial_\xi\ \psi_2\right)\right)\ d\xi \\ \int_0^r \overline{\psi_2}\ .\ \left(\left(i\ c\ \hbar\ \partial_\xi\ \psi_1\right)\right)\ d\xi & \int_0^r \overline{\psi_2}\ .\ \left(\left(i\ c\ \hbar\ \partial_\xi\ \psi_2\right)\right)\ d\xi \end{pmatrix}\ \text{// MatrixForm}$$

Out[244]//MatrixForm=

$$\begin{pmatrix} -\dfrac{c\,\pi\,\hbar}{r} & 0 \\ 0 & \dfrac{c\,\pi\,\hbar}{r} \end{pmatrix}$$

In this basically one dimensional system the Hamiltonian matrix is diagonal.

$$\begin{pmatrix} -\dfrac{c\,\pi\,\hbar}{r} - \dfrac{c\,\pi\,\hbar}{n\,r} & 0 \\ 0 & \dfrac{c\,\pi\,\hbar}{r} + \dfrac{c\,\pi\,\hbar}{n\,r} \end{pmatrix}$$

The eigenvalues of the system are simply the diagonal entries.

$$\left\{ -\dfrac{c\,\pi\,\hbar}{r} - \dfrac{c\,\pi\,\hbar}{n\,r} , \dfrac{c\,\pi\,\hbar}{r} + \dfrac{c\,\pi\,\hbar}{n\,r} \right\}$$

The eigenvectors of this system are of very simple form.

$$\begin{pmatrix} 1 \\ 0 \end{pmatrix}, \begin{pmatrix} 0 \\ 1 \end{pmatrix}$$

Note how the Hamiltonian operator and the eigenvalues do not depend directly on the mass of the particles involved. These eigenvalues would be just as valid for completely massless particles as for massive particles. "n" in this formula would be the number of planck length's long the de Broglie wavelength of the particle is at the given height. Obviously the minimum wavelength would be determined by the "rest" or ground state wavelength of the particle. Notice that in this Hamiltonian operator explicit G has canceled out. Newtons gravitational constant is still a factor however. Keep in mind the postulate of quantized space time. Both the de Broglie wavelength and the altitude r would be integer multiples of the Planck length. The planck length is defined in terms of G.

Numerical analysis

Let me plug in some real numbers and see what size these energy levels are. For this formula I need the ground state deBroglie wavelength of the particle. In terms of planck length's this will have to be.

In[3]:=

$$ni = \left(500 \text{ kg} * \dfrac{1.616 * 10^{-35} \text{ meter}}{2.17665 * 10^{-8} \text{ kg}} \right) \dfrac{1}{1.616 * 10^{-35} \text{ meter}}$$

Out[3]=

$$2.29711 \times 10^{10}$$

In[1]:=

$$-\dfrac{\left(299\,792\,458\, \frac{\text{meter}}{\text{s}}\right) \pi \left(6.5821191556 * 10^{-16} \text{ ev s}\right)}{r} - \dfrac{\left(299\,792\,458\, \frac{\text{meter}}{\text{s}}\right) \pi \left(6.5821191556 * 10^{-16} \text{ ev s}\right)}{2.29711 \times 10^{10}\, r}$$

Out[1]=

$$-\dfrac{6.19921 \times 10^{-7} \text{ ev meter}}{r}$$

Plug in the altitude of the satellite when it is in geostationary orbit. This equation will tell us the energy in Joules needed to put the satellite into geostationary orbit. This does not take into account the launch vehicle, fuel weight etc. This would have to be the net energy imparted to the satellite.

In[16]:=

$$\left(\dfrac{6.19921 \times 10^{-7} \text{ ev meter}}{35\,786 \times 10^{3} \text{ meter}} - \dfrac{6.19921 \times 10^{-7} \text{ ev meter}}{1.61624 \times 10^{-35} \text{ meter}} \right) \dfrac{1}{\text{evj}}$$

Out[16]=

$$-6.14527 \times 10^{9} \text{ Joule}$$

The negative sign indicates that this is a binding energy. For comparison the classically computed energy is the Sum of it's kinetic and potential compontents.

$$\dfrac{1}{2} m v^2$$

In[92]:=

$$\frac{1}{2}\ (500\ \text{kg})\ \left(3.07 * 1000\ \frac{\text{meter}}{\text{s}}\right)^2$$

Out[92]=

$$\frac{2.35623 \times 10^9\ \text{kg}\,\text{meter}^2}{\text{s}^2}$$

$$\frac{\text{GMm}}{\text{r}}$$

In[31]:=

$$\frac{\left(6.67 \times 10^{-11}\right)\ \left(5.98 \times 10^{24}\right)\ 500}{\left(6.37 \times 10^6\right) + 35\,786 * 10^3}$$

Out[31]=

$$4.73083 \times 10^9$$

The sum of thses energies.

In[32]:=

$$2.35623 \times 10^9 + 4.73083 \times 10^9$$

Out[32]=

$$7.08706 \times 10^9$$

The percentage difference between my theories value and the classical value is.

In[35]:=

$$\frac{\left(6.14527 \times 10^9\right) - \left(7.08706 \times 10^9\right)}{\left(7.08706 \times 10^9\right)} * 100$$

Out[35]=

$$-13.2888$$

So the value computed from the theory of quantized space-time is 13.2% smaller than the classically computed value. For a more careful comparison I would need to solve the Full configuration interaction of the satellite with the earth in all four space-time dimensions using the theory of quantized space time. Then I would need to use the Schwarzschild metric to find the energy of the satellite. However I think that this analysis shows that my theory is consistent with established physics where it needs to be.

Conclusions and Summation.

What I have shown in this composition is that the theory of quantized space-time gives a simple method for solving for the quantum gravitational field. Simple basic matrix multiplication is all the math that is required. There is complexity in this theory in the form of some unfamiliar concepts. Concepts such as space-time being discrete and discontinuous. I know that the assumption that space is a featureless smooth background on which more interesting things occur is deeply ingrained. There is no fundamental reason to cling to that one piece of classical thinking. I know it will be hard for the reader to let go of such a pervasive and basic concept. However the results above speak for themselves.

I have shown in many ways that this theory gives results that are in accord with classical general relativity. I have shown that this theory gives results that have corrections at both the quantum and cosmological length scales.

In the section "Quantum Space-Time Dynamics" I have found the Lagrangian of the field that is invariant under the F(4)/Spin(4) symmetry. Moreover I have shown by induction that this Lagrangian is the correct Lagrangian for Quantum Space-Time dynamics.

I have shown that the conserved current in this theory is none other than the flow of space-time. The consequences of this are that space-time cannot be created or destroyed and that it is in finite supply. This result will have a great impact on cosmology and physics. This result troubled me at first as it seems counterintuitive to think that space-time is not simply a void into which our universe expanded after the big bang. In fact this conservation law says that the total amount of space-time that there is finite, and that space-time cannot be destroyed or created. As counter intuitive as it is on first sight on second sight it makes perfect sense. So far as I know there is no interaction that creates or destroys space-time. There are interactions that bend and fold it but none that create or destroy it. Therefore space-time itself must be a conserved quantity.

I have addressed the key question, the litmus test for all theories of quantum gravity, the thermodynamics of black holes. Formulae were presented for the volume, area, and Schwarzschild radius of a black hole. These formulae were used to find out how many particles could fit in the hole. The definition of entropy from information theory was then applied to the system. This gave a formula for the incremental change in entropy. An argument was presented for the use of the Maxwell-Boltzman statistics on the particles in the hole based on the fact that their identity as boseons or fermions cannot be determined and the number of particles bound to the hole is so low that their are many more states than particles. With that distribution settled upon to my satisfaction the instantaneous entropy of the black hole is computed by way of Boltzmans formula. Then using the first law of thermodynamics the temperatures of a Schwarzschild then a Kerr-Newmann black hole are written down.

These formulae provide a mathematical model for gravitational collapse which is quantum mechanically and relativistically acceptable. This is so because thermodydamic formulas do not depend on space-time coordinates and are all in terms of quantities which should be invariant under Lorentz, covariant transformations. Therefore this formulation is in agreement with special and general relativity as well as the theory of quantized space-time.

I have made concrete postictions of known gravitational experimental results.

I have managed to unify gravity with quantum mechanics not by "quantizeing general relativity" but by comming up with a theory that addressed the specific problems of geometry a the smallest length scale at which "geometery" would make sense.

References

1 "Spin Foam Model for Quantum Gravity", A.Perez, arXiv:gr-qc 0301113 v2 14 feb, 2003

2 "Loop and Spin Foam Quantum Gravity: A Brief guide for Beginners", H Nicolai and K, Peeters, arXiv:hep-th/0601129 v1.

3. "M(atrix) Thory: Matrix Quantum Mechanics as a Fundamental Theory", W. Taylor, arXiv:hep-th/0101126 v2 2 Feb 2001.

4."Blackholes and information theory" J.D. Beckenstien, arXiv:quant-ph/0311049 v1 9 Nov 2003

5."Discrete Lorentzian Quantum Gravity", R. Loll, arXiv:hep- th/0011194 v1 21 Nov 2000.

6."Gravitomagnetic effects", M.L. Ruggiero, A. Tartaglia, arXiv:gr- qc/0207065 v2 23 Jul 2002

7. The Principle of Relativity ,("The foundation of the General Theory of Relativity" A. Einstein, Annalen der Physik, 1916.) Dover Publications, inc, New York, 1952

8."Measurement of quantum states of neutrons in the Earth's gravitational field", Nesvizhevsky et al. , Phys. Rev. D 67, 102002 (2003) [9 pages]